GRAVITY, PARTICLES, AND ASTROPHYSICS

ASTROPHYSICS AND SPACE SCIENCE LIBRARY

A SERIES OF BOOKS ON THE RECENT DEVELOPMENTS
OF SPACE SCIENCE AND OF GENERAL GEOPHYSICS AND ASTROPHYSICS
PUBLISHED IN CONNECTION WITH THE JOURNAL
SPACE SCIENCE REVIEWS

Editorial Board

J. E. BLAMONT, *Laboratoire d'Aeronomie, Verrières, France*

R. L. F. BOYD, *University College, London, England*

L. GOLDBERG, *Kitt Peak National Observatory, Tucson, Ariz., U.S.A.*

C. DE JAGER, *University of Utrecht, The Netherlands*

Z. KOPAL, *University of Manchester, England*

G. H. LUDWIG, *NOAA, National Environmental Satellite Service, Suitland, Md., U.S.A.*

R. LÜST, *President Max-Planck-Gesellschaft zur Förderung der Wissenschaften, München, F.R.G.*

B. M. McCORMAC, *Lockheed Palo Alto Research Laboratory, Palo Alto, Calif., U.S.A.*

H. E. NEWELL, *Alexandria, Va., U.S.A.*

L. I. SEDOV, *Academy of Sciences of the U.S.S.R., Moscow, U.S.S.R.*

Z. ŠVESTKA, *University of Utrecht, The Netherlands*

VOLUME 79

PAUL S. WESSON

*St. John's College, Cambridge University, England and
Institute for Theoretical Astrophysics, Oslo University, Norway*

GRAVITY, PARTICLES, AND ASTROPHYSICS

*A Review of Modern Theories of Gravity and G-variability,
and their Relation to Elementary Particle Physics and
Astrophysics*

D. REIDEL PUBLISHING COMPANY

DORDRECHT : HOLLAND / BOSTON : U.S.A.
LONDON : ENGLAND

Library of Congress Cataloging in Publication Data

Wesson, Paul S
   Gravity, particles, and astrophysics.

   (Astrophysics and space science library ; v. 79)
   Bibliography: p.
   Includes index.
      1. Gravity.   2. Particles (Nuclear physics).
   3. Astrophysics.   I. Title.   II. Series.
   QB331.W47   531'.14   80–14820
   ISBN 90–277–1083–X

Published by D. Reidel Publishing Company,
P.O. Box 17, 3300 AA Dordrecht, Holland.

Sold and distributed in the U.S.A. and Canada
by Kluwer Boston Inc., Lincoln Building,
160 Old Derby Street, Hingham, MA 02043, U.S.A.

In all other countries, sold and distributed
by Kluwer Academic Publishers Group,
P.O. Box 322, 3300 AH Dordrecht, Holland.

D. Reidel Publishing Company is a member of the Kluwer Group.

All Rights Reserved
Copyright © 1980 by D. Reidel Publishing Company, Dordrecht, Holland.
No part of the material protected by this copyright notice may be reproduced or
utilised in any form or by any means, electronic or mechanical
including photocopying, recording or by any informational storage and
retrieval system, without written permission from the copyright owner

Printed in The Netherlands

PREFACE

This book deals with the relationship between gravitation and elementary particle physics, and the implications of these subjects for astrophysics. There has, in recent years, been renewed interest in theories that connect up gravitation and particle physics, and in the astrophysical consequences of such theories. Some of these accounts involve a time-variation of the Newtonian gravitational parameter, G. In this respect, the present book may be regarded as a companion to my *Cosmology and Geophysics* (Hilger, Bristol, 1978). There is some overlap as regards the discussion of G-variability, but the emphasis in the present book is on astrophysics while the emphasis in the other one is on geophysics.

The subject is a very broad one indeed, and in giving a review of it I have adopted a somewhat unorthodox way of presenting the material involved. The main reason for this is that a review of such a wide subject should aim at two levels: the level of the person who is interested in it, and the level of the person who is professionally engaged in research into it. To achieve such a two-level coverage, I have split the text up into two parts. The first part (Chapters 1–7) represents a relatively non-technical overview of the subject, while the second part (Chapters 8–11) represents a technical examination of the most important aspects of non-Einsteinian gravitational theory and its relation to astrophysics. These two parts of the book are of comparable lengths. I would like to emphasize here, as I have done in the text, that the second part of this book is of equal standing with the first part.

A review, to be effective, has to be comprehensive. I have included the relevant references from the fields of gravitation, elementary particle theory and astrophysics. This is a large territory to cover, and that is why there are a lot of references. I hope the bibliography will be of assistance to the professional researcher. In the text, the symbol ($\simeq$) means equal to within a factor of three, ($\approx$) means equal to within an order of magnitude, while ($\sim$) means expected to be asymptotically equal.

I would like to thank the following people who have, either directly or indirectly, been of assistance in the preparation of this book: V. Canuto, A. Lermann, D. Lynden-Bell, I. Roxburgh, R. Stabell and S. J. Wesson. However, the responsibility for any omissions or errors rests with me.

PAUL S. WESSON

CONTENTS

PART ONE

PREFACE	v
CHAPTER 1. INTRODUCTION	3
CHAPTER 2. VARIABLE-G GRAVITATION	5
2.1. Introduction	5
2.2. Three Variable-G Theories	9
2.3. The Status of Variable-G Gravitation	28
CHAPTER 3. PARTICLE PHYSICS AND GRAVITATION	31
3.1. Introduction	31
3.2. Gauge Theories and Gravitation	32
3.3. Particle Physics and Cosmology	37
CHAPTER 4. ASTROPHYSICS	41
CHAPTER 5. GEOPHYSICS	48
CHAPTER 6. DISCUSSION	57
6.1. Introduction	57
6.2. Gravity Experiments	58
6.3. Scale Invariance and Self-Similarity	66
CHAPTER 7. SUMMARY	81

PART TWO

CHAPTER 8. SCALE INVARIANCE AND SCALAR-TENSOR THEORIES	89
CHAPTER 9. ALTERNATIVE THEORIES OF GRAVITY	93
9.1. Introduction	93

9.2. A Compendium of Gravitational Theories	93
9.3. An Intercomparison of Gravitational Theories	112
CHAPTER 10. GROUP THEORY AND GRAVITY	**114**
10.1. Introduction	114
10.2. The Conformal Group and Related Topics	114
10.3. Other Aspects of Group Theory	121
CHAPTER 11. THE STATUS OF NON-DOPPLER REDSHIFTS IN ASTROPHYSICS	**135**
11.1. Introduction	135
11.2. The Tired-Light and Related Theories	135
11.3. Redshift Anomalies in Astronomical Sources	143
11.4. Non-Friedmannian Redshifts	147
REFERENCES	**162**
INDEX	**187**

PART ONE

CHAPTER 1

INTRODUCTION

Gravitation is a subject which on the theoretical side has been characterized by two successes and innumerable failures. The two successes are Newton's and Einstein's theories of gravity. The failures are all the other theories which have not passed the critical test of comparison with observation and have so been discarded. Recent progress in the search for a wider theory of gravitation than Einstein's has been noticeably influenced by the belief that gravitation and elementary particle physics may not be completely separate disciplines, but that theories of matter on the large and small scales may both be manifestations of a more fundamental theory that underlies both subjects. The most discussed way in which the unification of these two areas of physics might be brought about concerns a possible variability of G, the gravitational parameter.

The Newtonian gravitational parameter G is a constant in Newton's law of gravitation (force = $-GM_1M_2/r_{12}^2$ where r_{12} is the separation of two bodies of masses M_1 and M_2), and in Einstein's general theory of relativity. However, over the last 50 years there have been numerous suggestions that G might in fact be variable. The detailed grounds for these suggestions have been different, but most G-variable theories are accounts of gravitation which have a cosmological basis. As such, they usually involve a time-dependent G: $G = G(t)$, where t is a parameter that can, loosely-speaking, be interpreted as the 'age' of the Universe. These G-variable theories have consequences that are expected to show themselves most clearly in the fields of astrophysics and geophysics, since it is on the large scale that gravitation is dominant over other forces of Nature.

The purpose of the present account is twofold. Firstly, to review the variable-G concept as it relates to astrophysics and geophysics. A comprehensive technical review of this subject is already available which examines the status of work published up to the beginning of 1977 and which concentrates on geophysics (Wesson, 1978a). The present account therefore concentrates on results which have appeared during the last few years and on astrophysics. Secondly, the large amount of effort which is being expended on all aspects of variable-G cosmology makes it desirable to isolate the main outstanding problems of this subject.

A list of problems that need clarification is given below (Chapter 7), after an account of variable-G gravitation (Chapter 2), its possible connection with elementary particle theory (Chapter 3), and its implications for astrophysics (Chapter 4) and geophysics (Chapter 5). A general discussion of these and related topics is given in Chapter 6. Four technical chapters (8–11) discuss respectively the status of scalar-tensor variable-G theory, alternative theories of gravity, group theory and gravity, and non-Doppler redshifts in astrophysics. The last four chapters are not intended to be repositories for information that is of lesser interest than that contained in the first seven. Rather, the intention is that the second part of the book will be of interest mainly to those doing research in gravitation and astrophysics, while the first part is designed for those whose primary objective is to gain a simple overview of the field. By adopting this binary approach, it is hoped that the present account will have the widest possible usefulness both as a review and as an introduction to the subject.

CHAPTER 2

VARIABLE-G GRAVITATION

2.1. Introduction

Numerous theories have been proposed in which the Newtonian gravitational parameter G is variable, but progress in astrophysics and geophysics has winnowed out the field somewhat so that today there remain only three main competitors: (a) Dirac's theory (Dirac, 1938, 1973); (b) the Hoyle/Narlikar theory (Hoyle and Narlikar, 1971a); and (c) the scale-covariant theory of Canuto et al. (1977a). In former years, much effort was expended on testing the consequences of the Brans/Dicke scalar-tensor theory (Brans and Dicke, 1961) but research in solar physics has shown that even if the theory is correct then it does not differ significantly from Einstein's general relativity in its consequences (Chapter 8; Wesson, 1978a). Scepticism about the presence of a scalar interaction in gravitation has led also to a feeling that it is unlikely that there could be a significant vector interaction present either, so while it may be mathematically viable a vector-tensor theory of gravitation (Hellings and Nordtvedt, 1973; Ni, 1972) is not favoured as a possible alternative to the (pure tensor) theory of general relativity.

This leaves the three theories (a), (b), (c) as the main contenders for a possible new theory of gravitation that might replace Einstein's theory. The latter is still taken as a reference theory in cosmology, and predictions of (a), (b), (c) are often presented in terms of expected departures from general relativity. However, general relativity and the new theories have something in common: they are all metric theories. That is, they are based on the use of an interval ds which is an element of distance between two points in four-dimensional (three space directions plus time) space-time. The square of ds is written as

$$ds^2 = g_{ij}\, dx^i\, dx^j,$$

where dx^i and dx^j are coordinate increments (i.e., dx, dy, dz, and $c\,dt$ in simple rectangular coordinates) and g_{ij} are dimensionless functions of the coordinates that need to be found from the field equations of a given theory. The interval is an abstract quantity, a kind of shorthand way of describing how the matter in the Universe affects the propagation of light and the motions of particles in the space-time. The real physics of a metric theory is contained in the field equations. For example, the field equations for Einstein's theory in problems to do with spherically-symmetric distributions of matter (i.e., *most* problems) can be expressed as a set of coupled, partial differential equations. These equations connect the metric coefficients g_{ij} (which describe how ds depends on the coordinates x_1, x_2, x_3, ct) with the properties of the matter: the pressure (p), density (ρ), and mass (m) within some three-dimensional distance $r = (x_1^2 + x_2^2 + x_3^2)^{1/2}$ from a chosen origin of coordinates. Three constants of physics are also involved: c (the velocity of light), G (the Newtonian gravitational parameter) and Λ (the cosmological constant). The last represents a force that tries to push matter apart if $\Lambda > 0$ and pull matter together if $\Lambda < 0$. It can also be thought of as an invariable material background (see below). The value of Λ is small, and it might even be zero, but the question of its actual size is as yet undecided.

In Einstein's general relativity, c, G and Λ are all proper constants. In particular, the value of G has to be a constant since G-constancy is inbuilt as a manifestation of the Principle of Equivalence. Einstein's Principle of Equivalence states that all local, freely falling nonrotating laboratories are fully equivalent for the performance of all physical experiments. This form of the Principle is due to Rindler (1969). The restriction to nonrotating frames avoids complications which arise when one considers the possible effect of rotating distant matter in inducing inertia in nearby matter. Such a possibility is connected with Mach's Principle, which is usually taken to mean that the properties of local matter (e.g., its inertia) are connected with the properties of the rest of the matter in the Universe. This latter principle is widely regarded as the second foundation stone of general relativity (after the Principle of Equivalence). But although it has some justification in terms of studies such as those to do with the well-known massive rotating shell model of Thirring (1918, 1921; see also Lense and Thirring, 1918) and a related problem in electrostatics (Schiff, 1939), the status of Mach's Principle in relation to Einstein's theory is ambiguous (Rindler, 1969; Wesson, 1978a), and will not be discussed further. We shall concentrate here on the Principle of Equivalence, the most important consequence of which is that the dynamical effect of a gravitational field and an acceleration are equivalent.

The Principle of Equivalence was propounded as a basis for Einstein's theory, but in connection with metric theories of gravitation in general it is usually discussed in one of two forms: (i) the strong form, on which Einstein's theory of relativity is based, is essentially that quoted above, and implicitly includes a statement that the laws of physics and the numerical values of the constants that enter into them (e.g., G) are the same at all places and at all times; (ii) the weak form says that in a given gravitational field all bodies experience the same acceleration, which can also be shown to be the same as saying that the mass which gives rise to a gravitational field (the gravitational mass) is the same as that which responds to a force in terms of an acceleration (the inertial mass). Actually, one can also distinguish between active and passive gravitational masses as those giving rise to and responding to a gravitational field respectively, but the concept of passive mass is basically Newtonian and does not play an essential role in relativistic gravitational theory. The strong form of the Principle of Equivalence has not been extensively tested, and the theories in which G is variable represent examples where the strong form of the Principle of Equivalence is denied. The main reason why the strong form of the Principle of Equivalence is highly regarded is that it represents a very logical and therefore compelling foundation for gravitational theory. The weak form of the Principle of Equivalence is much better supported by experiment, and it is in fact this form of the Principle which is usually implied by the phrase 'Principle of Equivalence'. In this form, it is supported by various modifications of the Eötvös experiment (Wesson, 1978a), which establish the identity of gravitational and inertial mass to high accuracy, and most workers regard the Principle of Equivalence as being well established both experimentally and theoretically (Wapstra and Nijgh, 1955; Roll et al., 1964; Braginskii and Panov, 1972; Rindler, 1969; Weinberg, 1972, pp. 67–90; Wesson, 1975a, 1978a). It is instructive to take a brief look at some of the evidence involved before proceeding to consider theories which break with the strong form of the Principle if not with the weak form.

The Principle of Equivalence is supported by the observation that the ratio of (gravitational)/(inertial) masses is the same for neutrons and (protons plus electrons) for a range of different substances including glass, cork, antimonite and brass, the noted ratio being unity to within 1 part in 6×10^5 (Wapstra and Nijgh, 1955). Indeed this ratio is known to be unity to great accuracy for certain substances. For aluminium and gold the ratio is unity to 3 parts in 10^{11} (Roll et al., 1964), while for aluminium and platinum it is unity to 1 part in 10^{12} (Braginskii and Panov, 1972). While these accuracies are impressive, it must be realized that they concern bodies of sizes that can be conveniently handled in the laboratory. The Principle of Equivalence for more massive bodies (e.g., planets and stars) has been much discussed (see, e.g., Nordtvedt, 1968, 1969, 1971; Wesson, 1978a) and is not observationally so well established as it is for objects with which we are familiar in the laboratory. For bodies of planetary size, laser tracking of the Moon in its orbit around the Earth shows that gravitational binding energy contributes equally to the Earth's inertial and gravitational masses to within 1.5–3.0 percent (Shapiro et al., 1976; Williams et al., 1976). However, for bodies of stellar size, and especially for collapsed objects such as neutron stars, the Principle of Equivalence has not been rigorously tested. One feasible way of testing the Principle of Equivalence in its strong form (meaning that the laws of physics are the same at all places and at all times and are not affected by the strength of the gravitational field) has been suggested (Brecher, 1978). The proposed test would involve observations of X-ray and γ-ray lines produced by different physical processes near the surface of a neutron star. This test has not yet been carried out, and likewise a possible breakdown of the Principle of Equivalence for charged bodies falling in a gravitational field (McGruder, 1978; see also Nordtvedt, 1970a) has yet to be tested experimentally.

A breakdown of the Principle of Equivalence in any form, if detected, would constitute a departure from Einstein's general relativity. Possible departures depend, as far as the theories (a), (b), (c) are concerned, on the relevant field equations. But while the detailed bases for the three noted theories may be different, they share the property that their departures from Einstein's theory can be expressed in terms of G-variability. This is itself a consequence of the fact that in these theories there is a choice available concerning which time scale one should use to describe physical processes.

The meaning of the symbol for time which one sees in the equations of physics was discussed at a fundamental level by Milne and Whitrow (1938) and Milne (1940). There was some controversy about the reformulation of special relativity in a cosmological context which Milne termed kinematic relativity (Milne, 1933a, 1935, 1940, 1941, 1948; Milne and Whitrow, 1938; McVittie, 1940, 1941, 1942; Walker, 1941). The status of that theory is discussed in Wesson (1978a) and in Chapter 10. But irrespective of the standing of that theory as a cosmology, Milne pointed out the important fact that the apparent form of our laws of physics depends on how our clocks run. In particular, the velocities ($\dot{r}$) of the galaxies and their density distribution (ρ) can only be sensibly discussed within the confines of a set-up in which a local clock and the clocks on other galaxies have been graduated with respect to each other in a specific (but freely choosable) way. (In theory, the most simple type of clock in astronomy would be the photon clock, in which light signals bounce between two particles moving on parallel world lines (Harvey, 1976). In practice, clocks in astronomy are based on periodic processes in atoms, which by the use of various technical devices provide a domestic time service (Smith, 1969; Hellwig et al., 1978; Kartaschoff, 1978). The establishing of a 'uniform' time for use in astronomy has been discussed by Winkler and Van Flandern (1977); but it should be emphasized (Roxburgh, 1977a) that there is no absolute definition of a 'uniform' time, and it is impossible to construct a 'uniformly' running clock since the concept of 'uniform' is entirely a matter of convention.) There are two natural choices of time scale in physics: t-time or atomic time is the time which atomic and electromagnetic processes keep; τ-time or gravitational time is the time which large-scale, gravitationally-dominated systems keep. The relation between t and τ depends on the particular model one adopts of the Universe, but (a), (b), (c) have in common the condition that G can be variable as measured by a t-time clock (e.g., a quartz-crystal wristwatch). On the other hand, G is a true constant as measured by a τ-time clock (e.g., the motions of the planets). The difference, $G(t)$ = variable versus $G(\tau)$ = constant, is one way of formulating the mathematically complicated theories (a), (b), (c) in an easily-appreciated manner.

2.2. Three Variable-G Theories

(a) Dirac's theory has as its basis the Large Numbers Hypothesis (LNH). Dirac (1938, 1972) noticed that the ratio of electrical $(e^2/4\pi\epsilon_0 r^2)$ to gravitational $(Gm_p m_e/r^2)$ force between the proton (mass m_p) and electron (mass m_e) in a hydrogen atom is a large number of order 10^{40}: $e^2/4\pi\epsilon_0 G m_p m_e \approx 10^{40}$. (Here, e is the electron charge and ϵ_0 is the permittivity of free space in mks units. These expressions take a slightly simpler form in cgs units, where $e^2/4\pi\epsilon_0$ is replaced by e^2.) Similarly, the ratio of the present 'age of the Universe' $t = t_0 \approx 2 \times 10^{10}$ yr to the atomic unit of time $(e^2/4\pi\epsilon_0 m_e c^3)$ is of roughly the same size. Dirac suggested that the two numbers are in fact equal:

$$\frac{e^2}{4\pi\epsilon_0 G m_p m_e} \approx 10^{40} \approx \frac{4\pi\epsilon_0 m_e c^3 t_0}{e^2}. \tag{2.1}$$

Assuming that the atomic parameters do not vary with time, the last equation says that for general t

$$G \propto \frac{1}{t}, \tag{2.2}$$

which leads one to expect that $\dot{G}/G \approx -6 \times 10^{-11}$ yr^{-1} at the present epoch. This derivation of G-variability as judged by clocks keeping (atomic) t-time is a direct result of the LNH, which says that the two numbers in (2.1) should be equal, and (in generalized form) that dimensionless numbers of size 10^{40n} should vary with the epoch of the Universe t as t^n. The simple account of G-variability expressed in (2.1) has been considerably extended by Dirac (1973) and others using a reformulation of the metric and geometry of space-time that was originally due to Weyl.

The Weyl geometry (Weyl, 1922) is of interest because it is a natural framework for a scale invariant theory of gravity. A space-time is scale invariant if, under a change of the scale or standard of length by a factor $\beta(x)$ which may depend on the coordinates, the physical laws of the space-time remain unaltered. A change in the scale or standard of length in a space-time involves a change in the line element from ds to ds' where d$s' = \beta$ ds, and a corresponding change in the metric from g_{ij} to g_{ij}' where

$$g_{ij}'(x) = \beta^2 g_{ij}(x). \tag{2.3}$$

The metric coefficients g_{ij} in gravitation theory are analogues of the potentials of electromagnetic theory, which latter is invariant under changes of gauge (see Chapter 3). In gravitation theory, the function β is usually referred to as the gauge function, and a transformation of the type just discussed is called a scale or gauge transformation (in the form of (2.3) it is often also referred to as a conformal transformation). Invariance under changes of gauge is of established significance in electromagnetism, while invariance under changes of scale appears to represent an asymptotic symmetry in high-energy particle physics (Callan *et al.*, 1970; Anderson, 1971). It is therefore natural that attempts at constructing a gauge invariant theory of gravity have focussed on the use of the Weyl geometry.

Weyl's geometry was originally proposed as a basis for a unified theory of gravity and electromagnetism that was required to leave physical laws invariant under both arbitrary coordinate transformations and gauge or scale transformations. While the theory itself proved untenable, the geometry on which it was based is still viable. (See Anderson (1971) for a review of the Weyl theory and a discussion of scale invariance in relation to the Brans/Dicke scalar-tensor theory of gravity; and Weinberg (1972) for comments on gauge invariance in electromagnetism and general relativity. The latter is gauge invariant for small perturbations (Weinberg, 1972, pp. 252–254, 290–291), meaning that given a solution to Einstein's equations with metric g_{ij} it is always possible to obtain other solutions with slightly different forms for g_{ij} that have the same physical content.) There has been considerable work done on ways in which metric theories of gravity – like Einstein's theory and Dirac's theory – might be unified with electromagnetism and gauge theories of the elementary particles. One promising way of doing this is by employing the dual status (force per unit mass $\Lambda r/3$, and background density $\Lambda/8\pi G$ and pressure $-\Lambda c^2/8\pi G$) of the cosmological constant Λ (see Chapter 3 and Wesson, 1978a). The Dirac theory is compatible with gauge invariance in the Weyl sense, and one might expect that the symmetry breaking which is observed in particle physics could be explained in terms of some cosmological effect that couples to quantum-mechanical processes (Dirac, 1973, 1974; Wesson, 1978a). However, the status of symmetry breaking in Dirac's theory is doubtful because the field equation which involves the gauge function $\beta(x)$ can be decoupled from the equations which describe the matter (Pietenpol et al., 1974), suggesting that the implied cosmological symmetry breaking is actually unobservable.

Nevertheless, there are grounds for hoping that a gravitational theory which is based on the Weyl geometry and which incorporates a Λ term might in some way be compatible with elementary particle theory. The Dirac theory as it is based on the LNH is characterized by two natural gauges, in one of which $\Lambda \equiv 0$ while in the other Λ is finite. The LNH and the field equations of the theory in both gauges imply continuous creation of matter: this occurs as additive (+) creation (zero Λ gauge) or as multiplicative (×) creation (finite Λ gauge). In the first case, continuous creation occurs at the same rate everywhere in the Universe, while in the second case matter is created where matter is already most dense. It is interesting from the point of view of comparison with gauge theories of the elementary particles that most astrophysical tests which have been carried out using Dirac's theory have shown more or less conflict with the (+)-model and agreement (or at least, lack of conflict) with the (×)-model Wesson, 1978a). Physically, the reason for this is that in the (+)-model, the main consequence of the Dirac theory is that $G \propto t^{-1}$ while the continuous creation process does not noticeably affect the masses of astronomical bodies since most of the new matter is created in interstellar space. Many systems in astrophysics would be seriously affected by uncompensated changes in G of this type. For example, homology relationships in the theory of stellar structure lead one to expect that the luminosity of an average star depends on G and the mass of the star M_* as $L \propto G^7 M_*^5$ (Teller, 1948; Dicke, 1962a). Clearly, stellar evolution is sensitive to changes in G, and this has been the main method of testing variable-G theories. In comparison, for the (×)-model the continuous creation of matter where matter is already most dense causes the mass of the star to grow as $M_* \propto t^2$ where t is atomic time, and this process tends somewhat to offset the $G \propto t^{-1}$ dependency which is also present.

A detailed series of calculations on stellar structure and variable-G theories has been carried out by Maeder, with especial attention to theories (like Dirac's) that use the Weyl geometry as a metric background. The basis of these calculations are computer programs that follow the evolution of stars near the main sequence (Maeder, 1974), and from the main sequence to the red giant branch (Maeder, 1975). Programs for population I main sequence stars (Maeder, 1976) like the Sun, can already mimic the life history of a typical star in a way that agrees with observation *before* any G and M_*

variability is introduced. The test of variable-G theories is therefore to see if the programs yield stellar models that are in agreement with observation when the $G \propto t^{-1}$ and $M_* \propto t^2$ effects are added. Four separate things can be examined which show that the (+)-model ($G \propto t^{-1}$, $M_* =$ constant) must be rejected whereas the ($\times$)-model ($G \propto t^{-1}$, $M_* \propto t^2$) is acceptable (Maeder, 1977a). These things are: (i) the luminosity L of a star like the Sun as a function of time; (ii) the Hertzsprung/Russell diagram for a population of stars with ages and luminosities affected by G and M_* variability; (iii) the solar neutrino emission, which is decreased by a factor 2 in the ($\times$)-model as compared to conventional stellar evolution calculations, this being important in order to help explain the observed low rate of neutrino emission from the Sun; (iv) the flux of energy reaching the Earth from the Sun, this conflicting with geological data on the (+)-model, whereas on the ($\times$)-model it is approximately the same as with conventional ($G =$ constant, $M_* =$ constant) theory. These results show a strong preference for the ($\times$)-model over the (+)-model as far as stellar evolution calculations are concerned.

Once one knows how the luminosity of stars vary in the G-variable and G- and M_*-variable models, one can calculate how the brightness of a galaxy composed of stars varies as a function of t, or equivalently as a function of redshift z. A study of the Hubble/Sandage diagram for first-ranked galaxies with evolutionary corrections for the effects of Dirac's cosmology has been carried out (Maeder, 1977b). It shows that the simple $G \propto t^{-1}$ model can be ruled out (galaxies at high z would be too bright), while the $G \propto t^{-1}$, $M_* \propto t^2$ model is acceptable.

In a general astrophysical context, the use of Weyl's geometry alters slightly the equation of motion of an astronomical body in an expanding Universe (Maeder, 1978a). The acceleration $\ddot{r}$ is no longer given by Newton's law in its simple form but rather by

$$\ddot{r} = -\frac{GM}{r^2} + H\dot{r}, \qquad (2.4)$$

where the extra term involves Hubble's parameter H. This new law of motion arises when one assumes that the gauge function in (2.3) is $\beta = \beta(t)$, this making $k = k(t)$ in the equation $\mathrm{d}l = lk_i \, \mathrm{d}x^i$ which gives the change in length ($\mathrm{d}l$) of a vector of length l which is moved from position x^i to position $x^i + \mathrm{d}x^i$ in the Weyl geometry. If there exists a choice of the general function $\beta(x)$ which reduces Weyl's geometry to the familiar one of Riemann (on which general relativity is based), then the k_i are given by $k_i = -\partial_i(\ln \beta)$, where one recalls that in the Weyl geometry a scale or gauge transformation is defined by $\mathrm{d}s' = \beta(x) \, \mathrm{d}s$, and represents a change of the standard of length in the space-time, this standard being possibly variable. Dirac's two special cases of the Weyl geometry correspond to $\beta \propto t$ and $\beta \propto t^{-1}$ for (+)-creation and (×)-creation respectively.

In the preferred (×)-model, the scale factor of the Universe (which describes how the distance between two galaxies changes with time) is $S(t) \propto t$. The Universe as described in t-time (atomic time) therefore expands in proportion to t, with G decreasing in inverse proportion to t, and there was a big bang at $t = 0$. (The redshift of the galaxies in t-time can be interpreted in terms of the Doppler effect due to the expansion.) The (×)-model can, though, be equivalently described in τ-time (gravitational or Einstein time), in which system the Universe is static, G = constant and the big bang 'occurred' at $\tau = -\infty$. (The redshift of the galaxies in this case can be understood as a gradual change in the rate of ticking of t-time clocks – meaning atoms – as judged by τ-time standards.) The S-function in t-time defines $H(\equiv \dot{S}/S = 1/t)$, and the equation of motion (2.4) in the Weyl geometry shows in effect that the expansion of the Universe couples to smaller scale systems and can affect the motions of their constituent particles. In particular, (2.4) can modify the dynamics of galaxies moving within the potential well of a cluster of galaxies and so perhaps help remove some of the virial discrepancy (missing mass problem) for these systems (Maeder, 1978a, b). The possibility of finding a dynamical solution to the missing mass problem also arises in Prokhovnik's theory (see Chapter 9) and has been discussed by Lewis (1976), while Shields (1978) has suggested a solution based on the stellar luminosity relation $L \propto G^7 M_*^5$ noted above. Shields notes that if G were higher in the early Universe, stars would have evolved more quickly than they now do, and such burnt-out stars may be present today as a population III of hyper-evolved dark objects (populations I and II being the more normal stars found respectively in the spiral arms and optical haloes of spiral galaxies). Such dark stars may form extended, nonluminous haloes around galaxies, providing some of the missing mass in clusters. A combination of this idea with the dynamical effect expected from the Weyl geometry could well remove all of the virial discrepancy in clusters of galaxies. On a smaller scale, the Weyl geometry can affect the oscillations of stars perpendicular to the Galactic plane as these move around the centre of the Milky Way (Magnenat et al., 1978). The Weyl geometry leads to a larger velocity dispersion for the old disk population of stars than for younger objects, this being a consequence of the new geometry that can be compared to observed stellar dynamics, although it is not such an important test of variable-G cosmology as the stellar evolution calculations outlined above.

It is clear that astrophysical tests which can be deduced from the Weyl geometry support Dirac's (×)-model quite well. This is not, however, to say that the tests outlined in this chapter support only the Dirac theory. It is possible in principle to formulate both the theory of (b) Hoyle/Narlikar and the theory of (c) Canuto *et al.*, in terms of Weyl's geometry (Bouvier and Maeder, 1978). On the other hand, it is not completely established that the Weyl geometry is a suitable basis for gravitation, as can be appreciated from a consideration of two kinds of criticism that have been made of the Dirac theory, one to do with the temperature of the Earth and the other to do with the 3 K microwave background radiation.

The temperature T_E of the Earth's surface varies with time in the Dirac theory (and in other variable-G theories), as a result of the decrease with epoch of the gravitational parameter and the operation of the process of continuous creation. Roxburgh (1976) criticised Dirac's theory on the ground that the expected behaviour of T_E with time is not in agreement with the moderate temperatures (similar to those of today) which can be deduced from the existence of life on the Earth 3×10^9yr ago. The (×)-model, although it is in agreement with other data, predicts that the Sun's luminosity should vary as $L_\odot \propto G^7 M_\odot^5 \propto t^3$. This increase of luminosity with time is offset to a certain degree by the change in distance of the Earth from the Sun due to G-variability and continuous creation. Roxburgh noted that this variation in orbital distance would depend on whether the newly-created matter appears at rest with respect to the local matter or at rest with respect to the cosmological background. In either case, he came to the conclusion that the temperature at the Earth's surface would have been too low on the (×)-model to be compatible with geological data. (The (+)-model is no better, since in that case $L_\odot \propto t^{-7}$ and T_E would have been too high in the past.) This problem can, as realized by Roxburgh (1976), be avoided by considering different dependencies of the parameters involved on time, a possibility which exists if one departs from the $S \propto t$ behaviour of the cosmological scale factor preferred by Dirac for the (×)-model. Furthermore, an error in Roxburgh's analysis has been pointed out by Canuto *et al.* (1977b; see also Roxburgh, 1977b), the correction of which tends to lessen the temperature problem. Canuto *et al.* noted that while $L_\odot \propto t^3$ holds in the (×)-model, the change in the orbital distance of the Earth from the Sun cannot be evaluated using the Newtonian equation of motion employed by Roxburgh. Rather, one must use equations which are valid in atomic units, and when this is done one finds that the solar luminosity and orbital radius effects offset each other to a larger degree than Roxburgh's calculations would suggest. The offsetting of the two effects is not complete, but the part which is left over is only a slow warming with time according to $T_E \propto t^{1/4}$, which is probably acceptable as far as geological evidence is concerned.

The 3 K microwave background has a black-body (Planckian) spectrum to a high degree of accuracy, and is conventionally taken to be an all-pervasive radiation field that was produced during a fireball stage through which the Universe passed long ago (see Chapters 3 and 11). A lot is known both observationally and theoretically about the microwave radiation field, and this knowledge has been used in two arguments that have been brought against the Large Numbers Hypothesis.

The first argument has been made by Mansfield (1976), and involves the following line of reasoning: The LNH in its general form predicts a variation of dimensionless numbers in proportion to a power of the time t, where the power depends on the size of the number concerned; for photons with the temperature $T \simeq 3$ K as observed for the microwave background, one can combine T with Boltzmann's constant k and a particle mass m to form the dimensionless number kT/mc^2; the size of this number leads one to expect that the temperature of the microwave radiation should vary as $t^{-1/3}$ or $t^{-1/4}$ depending on whether m is taken as the mass of the proton or the mass of the electron respectively, such dependencies holding both for the (+)-form and for the (×)-form of the Dirac theory (Dirac, 1975; Canuto and Lodenquai, 1977); but in the observed Universe, the matter and the radiation (in the form of photons of the 3 K field) appear to be decoupled, and on simple physical grounds this means that in Dirac cosmologies the temperature of the photons should vary as $T \propto t^{-1}$ (Dirac, 1975; Mansfield, 1976); in order to (partially) resolve this discrepancy in temperature dependencies, Dirac has suggested that the 3 K background did not originate in a primeval fireball as assumed in conventional cosmology but instead emerged from a decoupling of matter and radiation from an intergalactic medium at a relatively recent epoch corresponding to redshifts (z) of about unity (Dirac, 1975; Mansfield, 1976); this suggestion matches the time dependencies of the radiation temperature for most of the redshift history of the Universe because prior to decoupling the radiation and the intergalactic medium would have been in equilibrium, the temperature of both photons and plasma varying in the same way and at a slower rate ($T \propto t^{-1/3}$ or $t^{-1/4}$) than after decoupling ($T \propto t^{-1}$; this way of expressing the problem is due to Mansfield (1976), and is equivalent to Dirac's formulation in which the mean free path of microwave photons in a cooling intergalactic medium is taken to be a distance corresponding to $z = 1$); the condition that the mean free path due to Compton scattering of a photon in the intergalactic plasma be equal to a distance corresponding to $z = 1$ can be used to estimate the density of the hypothetical intergalactic medium; this density turns out to be too high to allow radio waves from remote astronomical sources like QSOs to reach us without being absorbed by free-free interactions in the intergalactic medium. The implication of this line of reasoning is that the accurately Planckian form of the spectrum of the 3 K radiation may be compatible with its recent decoupling from an intergalactic medium, but that the latter is not compatible with radio astronomical observations. This apparently forces one to the conclusion that the intergalactic medium predicted by the LNH does not exist, and that therefore the LNH cannot be correct in its usually understood form.

However, this conclusion is not as damning as it appears because the line of reasoning employed by Mansfield involves the use of a parameter (T_0) which fixes the temperature of the hypothetical intergalactic medium at some chosen epoch, and the calculation of the optical depth to sources like QSOs is crucially dependent on this parameter. Canuto and Hsieh (1977) have pointed out that the actual intergalactic medium (if it exists) was probably reheated by astrophysical processes at some time in the past, and T_0 should really be considered a free parameter, rather than a quantity fixed solely by the cosmological model as assumed by Mansfield. This viewpoint is justifiable (even conventional cosmology has to make use of it in order to obtain low optical depths), and its incorporation into the calculation shows that the Dirac cosmology can be made consistent with radio astronomical observations of QSOs. But it must be emphasized that while the optical depth problem of the LNH can be resolved, the basic discrepancy noted above between the time dependence of the temperature of the 3 K radiation expected from the LNH ($T \propto t^{-1/3}$ or $t^{-1/4}$) and that which is believed to characterize the free (i.e., decoupled photons of the 3 K radiation ($T \propto t^{-1}$) still remains (Canuto and Lodenquai, 1977). This discrepancy represents a basic clash between the time-dependencies predicted by the numerological LNH and that predicted by the conventional physics of plasmas and gases. Dirac's hypothesis of recent decoupling reconciles the two dependencies for most of the redshift history of the Universe, but a full reconciliation on the basis of his cosmology can only be achieved if the 3 K radiation is still interacting with an intergalactic, ionized medium (or, equivalently, if the mean free path of photons in the intergalactic plasma is small, corresponding to redshifts $z \ll 1$). Such an interpretation would go directly against conventional views about the nature of the 3 K background. But perhaps the main problem with both the continual-interaction model and the $z \simeq 1$ decoupling model is that there is no evidence to support the existence of the required intergalactic medium.

The second argument against the LNH which uses knowledge about the 3 K microwave background has been made by Steigman (1978). In conventional cosmology, the precisely Planckian form of the spectrum of the microwave field is preserved as the Universe expands because the number of photons in a comoving volume element is conserved, (that is, photons are not created or destroyed to any significant degree). Steigman (1978) has criticized the Large Numbers Hypothesis and cosmologies compatible with it on the ground that the photon conservation condition would probably not be met in such theories, leading one to expect a large departure of the spectrum of the microwave background from the Planckian one which is observed. This criticism would seem to be valid and was also realized by Dirac (1975), who noted that creation of photons would lead to a departure from a black-body spectrum for the 3 K background radiation, the form of the departure depending on whether photon creation follows the rules of the (+)-model or the (×)-model. However, this argument against Dirac's theory would seem to hold only if (1) the continuous creation process of the LNH manifests itself as a creation of photons as well as matter, and if (2) the creation process occurs in intergalactic space. A form of the LNH that is confined to a creation of matter or to a creation of matter and radiation where matter is already most dense would seem able to circumvent Steigman's argument. If the LNH is applicable primarily to matter and not to radiation one can also avoid the cooling-rate discrepancy noted by Dirac (1975), Mansfield (1976), and Canuto and Lodenquai (1977) and which, as seen in the preceding paragraph, represents a problem for the Dirac theory in its usual form.

This proposal is not very congenial because it breaks with the fundamental nature of the Large Numbers Hypothesis. If one wishes to retain the basic nature of the LNH, and yet have agreement with data on the 3 K background, there are two things one can do: (i) The first is to keep the Dirac theory as it is based on the LNH but adopt a cosmological model different from those considered by Dirac. As far as the (×)-form of the theory is concerned, this means adopting a cosmology different from the $S \propto t$ model. A study of continuous-creation cosmologies which are consistent with the LNH has been given by Roxburgh (1977c). He has found a family of cosmological models which are in agreement with the validity of the LNH and the assumption that the equations of general relativity hold in one gauge (namely, that which uses Einstein units). Of the solutions in this family, two are the original $G \propto t^{-1}$ (no creation) and $G \propto t^{-1}$, (×)-creation models of Dirac, while another is a multiplicative steady-state model. The new models in the family are characterized by being spatially flat and having scale factors that vary as arbitrary powers of the epoch. However, the observational data which can be compared with the theoretical consequences of these models (e.g., their effects on the temperature of the Earth and the cosmological deceleration parameter) are not yet exact enough to enable a choice to be made from among the wide range of cosmologies that are compatible with the LNH. (ii) The second thing one can do in order to retain the fundamental nature of the LNH is to keep that hypothesis in its original form but change the nature of the underlying theory. This was done by Canuto *et al.* (1977a), who realized that the LNH could be used as a condition to choose the gauge most suited for describing physical phenomena in a theory that admits a range of gauges.

The theory of Canuto *et al.* (1977a) is examined in (c) below, but the meaning of the LNH has also been discussed in relation to the Dirac theory by Canuto and Lodenquai (1977). They have reviewed the status of Dirac's theory as it can be compared to observation, paying particular attention to aspects of the theory concerning the number/flux ratio for radio galaxies, pulsar spin-down and the variation with time of the Sun's luminosity. They conclude that the ($\times$)-form of the theory is in overall good agreement with observation. The only aspects of the LNH where difficulties arise concern the luminosities of white dwarfs (which are predicted on the basis of the LNH to be brighter than observed in practice) and the 3 K microwave background. The former difficulty concerns primarily the ($\times$)-form of the theory while the latter, as seen above, concerns both forms of the theory but is perhaps more serious for the ($+$)-form.

Thus, one sees that the ($\times$)-form of the Dirac theory, apart from a couple of oustanding problems that might be avoided by a change of astrophysical standpoint, is in agreement with observation. Further tests of the ($\times$)-model will probably concentrate on efforts to detect G-variability and continuous creation in the laboratory. Experiments to measure both of these effects are discussed in a general context in Chapter 6, but no results are available yet which conflict with the ($\times$)-model. Another way of detecting continuous creation which does not involve the building of elaborate experiments has been suggested by Gittus (1976). He has noted that the creation of atoms in the interstices of a crystalline structure would lead to an annealing out of dislocations in the material over a span of time, and that this effect sets a theoretical upper limit to the viscosity of crystalline materials. This upper limit to the crystal viscosity η is predicted to be $\eta = \mu/\phi$ where μ is the shear modulus of the material and ϕ is the creation rate of atoms in the sample in atoms per atom per second. A measurement of the ratio of the shear modulus to the crystal viscosity could thus give a direct indication of continuous creation of the type expected from the ($\times$)-form of Dirac's theory. The sensitivity of the method is expected to be such that it could detect continuous creation at a rate of 3×10^{-18} atoms per atom per second. This sensitivity is sufficient to detect creation at a rate of the cosmological order of $H \simeq 75$ km sec^{-1} Mpc^{-1} $\simeq 2.5 \times 10^{-18}$sec^{-1}, and the results of this and similar tests will clearly be of overriding importance for the future of Dirac's theory.

(b) The theory of Hoyle and Narlikar (1971a) is a theory of gravitation that is based on different principles to those of Dirac's theory, although the two theories have important technical aspects in common. A similar situation holds for the theory of Canuto *et al.* to be examined below, and since many of the comments made in connection with (a) above apply also to (b) and (c), accounts of the latter two theories will be restricted to those aspects in which they differ significantly from Dirac's theory.

There are two main things on which the Hoyle/Narlikar theory is based, namely conformal invariance and the absorber theory of radiation. The first is one of the aspects in which, technically, the Hoyle/Narlikar and Dirac theories are similar. The second is a theory of electromagnetism which, while discussed for some time in physics, had not been properly integrated into cosmology prior to the formulation of the Hoyle/Narlikar theory.

Conformal invariance means that the laws of physics (and the observable consequences of the laws as the latter are expressed in the form of differential equations), should be invariant under the carrying out of a space-time transformation of the type in which g_{ij} is changed to $g'_{ij} = \beta^2 g_{ij}$ as specified in (2.3). The importance of a conformal transformation is that the ratio of two lengths, possibly in different directions but at the same point, does not change under the transformation. The incorporation of this type of invariance into physics was justly regarded as being of fundamental importance by Hoyle and Narlikar. Maxwell's equations of electrodynamics are examples of equations of physics which possess the desired property of being conformally invariant. But Einstein's equations and the equations of classical particle theory are *not* conformally invariant. Hoyle and Narlikar therefore constructed a theory in which the equations governing gravitational phenomena are conformally invariant (i.e., in which all choices of the function $\beta(x)$ give physically equivalent results). Further, Dirac's equation in relativistic particle theory is made conformally invariant by introducing a variable particle mass $\beta^{-1} m_0$ in place of the conventional mass m_0 which is a constant.

The absorber theory of radiation is a kind of enlarged, symmetric theory of electrodynamics in which the response of the Universe in the future is regarded as being of equal standing with the response of the Universe in the past as regards observed electromagnetic interactions. Calculations in conventional electrodynamics use only retarded potentials, meaning potentials worked out at that time in the past when a given electromagnetic wave was emitted by a source. The absorber theory uses both retarded and advanced potentials, taking into account how the Universe will respond to an electromagnetic wave in the future. There are strong theoretical grounds for considering advanced potentials to have equal standing with retarded potentials (this is true at least mathematically), but observed electromagnetic processes only appear to involve retarded potentials. The absorber theory accounts for the apparent preponderance of effects due to retarded potentials in the observed world by assuming that the observable effects of the advanced potentials are effectively cancelled out due to an absorbing response of the Universe in the future to electromagnetic waves.

The two main principles on which the Hoyle/Narlikar theory is based have profound implications for astrophysics and geophysics. The consequences of the theory and their comparison with observation have been reviewed by Wesson (1978a). The most drastic departure concerns the variable nature of the masses of elementary particles and in particular of the electron mass m_e (see below). The behaviour of the mass is assumed to be given by a mass field $M(x)$, so that $m_e = \epsilon M(x)$ where ϵ is a dimensionless propagator. The mass field $M(x)$ is generated by the rest of the matter in the Universe, and its precise behaviour is linked to the cosmological solutions of the field equations of the theory (see Roxburgh (1977d) for a discussion of consistency conditions for the cosmological solutions of the mass integral formulation of the Hoyle/Narlikar theory and solutions of the corresponding formulation of general relativity). Spatial fluctuations in the mass field cause anomalous, gravitational redshifts, and in fact the evidence which had been adduced for the presence of non-Doppler redshifts in astronomical sources such as QSOs represented the original motivation for the proposal of the Hoyle/Narlikar theory. A discussion of the status of anomalous, non-Doppler redshifts in astrophysics is given in Chapter 11 (such redshifts are also predicted by theories other than that of Hoyle/Narlikar). It is clear from the evidence reviewed there that the case for non-Doppler redshifts is not now as strong as it was at the time the mass-field theory was proposed (Hoyle and Narlikar, 1971a). In fact, the case for the existence of such redshifts is decidedly weak.

Another type of criticism to which the theory is prone concerns astrophysical data connected with G-variability. The Hoyle/Narlikar theory is mathematically complex and does not have such a clear relationship to G-variability and continuous creation as that provided for the Dirac theory by the Large Numbers Hypothesis. Both processes can occur at arbitrary rates in the Hoyle/Narlikar theory, where they represent manifestations of the behaviour of the mass field and its propagator which can be described in cosmological terms by a certain choice of gauge function β. (Astrophysical problems can be evaluated in any convenient gauge, a property which is analagous to the choice of time scales available in the Dirac theory. Long-term changes in G are connected with long-term changes in particle masses and with the possible long-term variability of the mass field propagator by the condition $Gm^2 = 3\epsilon^2/4\pi$ which comes from the field equations.) A range of dependency of G on time t is allowed, but most attention has been focussed on a $G \propto t^{-1}$ dependency. However, G-variability in the Hoyle/Narlikar theory is not offset by continuous creation to the same degree as it is in Dirac's ($\times$)-model, and this leads to a rather poor agreement of the theory's predictions with observation in some astrophysical contexts. Most notably, the decrease of G with time means that stars and therefore galaxies should have been brighter and have had different colours in the past compared to now. This prediction can be tested by using observations of galaxies with high redshifts z: such sources lie at great distances, and so, because of the light-travel time from source to observer, can be viewed as they were at early epochs. For the reasonable case of G-variability according to $G \propto t^{-1} \propto (1 + z)^2$, a comparison of theory and observation (Barnothy and Tinsley, 1973) shows serious disagreement.

The negative comments of the two preceding paragrahs should not be taken necessarily as meaning that the Hoyle/Narlikar theory is not viable. The problems it faces in astrophysics can in principle be avoided by adopting parameters that bring its consequences more into line with those of conventional theory (i.e., general relativity). Further, the theory possesses some attractive astrophysical properties. For example, the (somewhat unpalatable) idea of a big-bang origin to the Universe is avoided in the Hoyle/Narlikar theory, where this singular event is replaced by a more natural one in which the propagator ϵ (which may be negative as well as being time-dependent) is taken to have passed through zero. Another attractive property is that the 3 K microwave background is adequately accounted for as being starlight which was thermalized at the time when ϵ passed through zero. This interpretation of the 3 K field is based on the fact that if ϵ passed through zero then the electron mass (which is $m_e = \epsilon M(x)$ in general), would also have passed through zero, causing the Thompson cross section ($8\pi e^4/3m_e^2 c^4$) to diverge and ensuring thermalization by scattering. This explanation of the thermal spectrum of the 3 K microwave background is superior to that which characterizes Dirac's theory.

Geophysically, the main consequences of the Hoyle/Narlikar theory involve an expansion of the Earth due to the effect of a decreasing G, and the change with time of the temperature T_E at the Earth's surface due ultimately to the same process of G-variability. Both these effects also enter into Dirac's theory, but the question of Earth expansion has not been examined in detail on the basis of that theory, while it has for the Hoyle/Narlikar theory. Earth expansion due to decreasing G has been discussed by Wesson (1978a), the amount of expansion expected on the basis of the Hoyle/Narlikar theory being about 10 km in the last 1×10^8 yr and 500 km over the whole history of the Earth, corresponding to a mean expansion rate of approximately 0.1 mm yr^{-1}. These figures depend somewhat on the rate of change of G, which on the Hoyle/Narlikar theory in its $G \propto t^{-1}$ form is $\dot{G}/G \simeq -1 \times 10^{-10}$ yr^{-1}. (For comparison, the Brans/Dicke theory mentioned in Chapter 1 involves G-variability at a rate of $\dot{G}/G \simeq -2 \times 10^{-11}$ yr^{-1} in its original form, with about 200 km of Earth expansion over the history of the Earth at a mean rate of approximately 0.05 mm yr^{-1}.) The amount of expansion which the Hoyle/Narlikar theory implies is compatible with geophysical data, although as will be seen in Chapter 5 there is evidence that some process other than G-variability may be contributing to the expansion of the Earth.

The effect of a variation in G on the luminosity of the Sun and so on the temperature of the Earth T_E is more drastic in the Hoyle/Narlikar theory than it is in the Dirac theory because the process involved is basically one of a variation in G uncompensated by significant continuous creation. Maiti (1978) has re-examined the variation in T_E due to changes in solar luminosity caused by changes in G. The relation due to Teller which gives the solar luminosity dependence as $L_\odot \propto G^7 M_\odot^5$ can be coupled with the expected variation in the Sun/Earth distance $r \propto (GM_\odot)^{-1}$, derived on the assumption that the gravitational force law has the quasi-Newtonian form of force $= -G(t)M_1 M_2/r_{12}^2$, to obtain the temperature dependence $T_E \propto (L_\odot/r^2)^{1/4} \propto G^{2.25} M_\odot^{1.75}$ (Teller, 1948; Maiti, 1978). It is obvious from the last relation that a variation of G with time that is not compensated by a notable change in $M_\odot$ with time must lead to very high values of T_E in the past. (There are reasons, based on the question of the consistency of a simple time-dependence of G in the Newtonian force law with the other laws of physics, for modifying the non-relativistic expression for the gravitational force. One such proposal has been made by Bishop and Landsberg (1976) and discussed by Wesson (1978a). Maiti has also examined the variation of T_E on the basis of the latter force law, finding $r \propto G/M_\odot$ in this case and $T_E \propto G^{1.25} M_\odot^{1.75}$, which is a less strong dependence than that given by the quasi-Newtonian force law.) The Hoyle/Narlikar theory leads one to expect that the temperature on the Earth's surface 3×10^9 yr ago was 70–80°C, from which high value it has decreased to the more comfortable one of the present day. However, the high temperature of the Earth in the past is not necessarily to be taken as being in contradiction with the development of life: primitive life forms (e.g., bacteria and algae) can live at high temperatures even though more advanced life forms cannot, and it has been argued that the development of life on the Earth from simple to complicated forms is evidence in favour of a decreasing temperature throughout geological time and so also of a decreasing value of G.

The preceding remarks concern that aspect of the Hoyle/Narlikar theory that involves the mass field and G-variability (see Hoyle, 1976, for a non-mathematical review). The other aspect of the theory, namely its incorporation of the absorber theory of electrodynamics, has been much less investigated, apparently because it is not a concept which is readily open to test. The quantum mechanical response of the Universe to electromagnetic waves has been discussed in relation to the absorber theory by Hoyle and Narlikar (1969, 1971b; see also Hoyle and Narlikar, 1968). As far as cosmology is concerned, there is little doubt that good arguments can be made for incorporating the perfect absorber condition into models of the Universe, but there is a large doubt that the actual Universe will satisfy the absorber criterion in the future. It is possible to formulate conditions for future opaqueness to electromagnetic radiation in certain classes of model Universe (see, e.g., Davies, 1973; Wesson, 1978a), but astrophysical data are not exact enough to decide if the actual Universe belongs to that group of models which will be perfectly absorbing.

The question of the applicability of the absorber theory is not completely academic, though. Partridge (1973) has reviewed the status of the theory and reported on an experiment in which he radiated microwaves alternately into free space and into a local absorber, hoping to detect advanced effects of an absorber theory of radiation as a difference in power drain between the two modes. He failed to detect any such effects. This does not necessarily mean that the absorber theory is wrong, because there is a doubt that the precise form of the experiment used by Partridge could detect the effects he was looking for (Pegg, 1973). It would in any case appear to be feasible to employ other experiments of related types in order to set constraints on the absorptive properties of the Universe, even if it is not clear to what extent such experiments will be able to decide if the absorber theory is valid or not. Although they are of doubtful significance as regards the absorptive properties of the Universe, Partridge's results are still of use. They have been used by Pegg (1977) in connection with a modified form of the absorber theory to put a limit on any *future* variation of Planck's constant using data which are available *now*. Pegg has concluded that $|\dot{h}/h| \leq 1 \times 10^{-18}$ yr^{-1} will hold in the future. This limit for a future variation of h is complemented by numerous experiments which have been made to measure a possible variation of h at the present epoch and in the past (Wesson, 1978a; Chapter 4). These experiments are astrophysical in nature and indicate that h has been varying at a rate of less than 5×10^{-13} yr^{-1} and probably at a rate of less than 2×10^{-14} yr^{-1} (Baum and Florentin-Nielsen, 1975, 1976). Such results, which show that h is not varying even on time scales longer than the conventional age of the Universe ($\approx 10^{10}$yr), do not, of course, represent criticism of the usual (h = constant) form of the absorber theory. The latter may well be viable, and the work of Partridge and Pegg shows that in principle some aspects of the absorber hypothesis can be tested even if it is not yet practicable to make any statement about its applicability to the real Universe.

In the absence of any definite evidence one way or the other about the acceptability of the absorber theory of electrodynamics, the consequences of the mass field concept must be employed to judge how far the Hoyle/Narlikar theory is in agreement with observation. As seen above, there is some astrophysical data against the viability of the theory in the usually-discussed form in which it leads to a fairly strong variation of G with time. But there is also some astrophysical data in favour of it, especially as regards providing a consistent interpretation of the 3 K microwave background. In other areas, the theory appears to be in acceptable agreement with observation. Its main claim to consideration as a workable theory of gravitation lies in the logicality of its foundations, while the main challenge it will have to meet in the future is whether the consequences of the principles on which it is based can survive comparison with more exact astrophysical data.

(c) The theory of Canuto *et al.* (1977a) is the most recent of the main variable-G theories, and although it has not been developed in as great detail as the preceding two theories, it is clear that it represents a viable account of gravitation in which G is time-dependent.

The theory is scale covariant (or scale invariant) and incorporates this property into a metric formulation by using the Weyl geometry discussed above. The gauge function β is arbitrary, but can be fixed by a choice of units. For example, for gravitational units, the gauge can be chosen so that the standard equations of Einstein's general relativity are recovered (these holding with G = constant and constant masses). Alternatively, for atomic units, the gauge would in general be expected to be different to that which is relevant to gravitational phenomena, and in such an atomic gauge both the value of G and the masses of bodies would vary with time. The choice of units and gauges which is present in the theory of Canuto *et al.* is analogous to that which is present in the theory of Dirac (although the theories are not equivalent), and Canuto *et al.* (1977a) suggested that the Large Numbers

Hypothesis should be used in their theory as a way of fixing the atomic gauge in relation to the gravitational gauge. The field equations of the theory (in a general gauge) are derived from a scale-invariant action principle, and an application of the equations of motion in atomic units to cosmology yields a family of homogeneous solutions having scale factors of the Dirac form $S \propto t$ for large cosmological times. This suggests that the LNH may be an asymptotic (large t) property of physics and not a hypothesis which holds exactly at all cosmological epochs.

The equations of motion can also be used to study Solar System dynamics, and the generalized hydrodynamical equations (derived for atomic units) can be used to derive the stellar structure equations which govern the evolution of stars like the Sun. Both the Solar System and stellar structure equations are similar to those of conventional theory except that they involve secular changes of some parameters on a time scale of the order of the age of the Universe. For the Sun, the secular effects involve variation with time of G and $M_\odot$, leading to a secular variation in the solar luminosity. The effects of such secular changes on the past climatology of the Earth and other aspects of geophysics were studied in preliminary form by Canuto et al. (1977a). They showed that the scale-covariant theory predicts Earth expansion at rates of approximately 0.02–0.03 mm yr^{-1} if there is no matter creation and 0.2–0.3 mm yr^{-1} if there is matter creation, the rate depending somewhat on the assumed age of the Universe ($t_0 = 1.5$–2.0×10^{10} yr). They also showed that the equation of conservation of angular momentum in the scale-covariant theory, which relates the spin-down of a rotating body with period P (rate $\dot{P}/P$) to changes in G (rate $\dot{G}/G$), β (rate $\dot{\beta}/\beta$) and the mass M (rate $\dot{M}/M$) predicts a rate of change of the Earth's period of rotation at the present epoch of approximately 0.06–0.6 msec/century, again depending on the question of creation and the assumed age of the Universe. Neither the expansion rates nor the spin-down rates predicted by the scale-covariant theory are in contradiction with observation, and Canuto et al. (1977a) concluded that their theory is in agreement with both cosmological and geophysical data.

Predictions of the scale-covariant theory have been examined in greater detail by Canuto and his coworkers, especially as the theory uses the LNH to fix the gauge function β in atomic units. The LNH says that $G \propto t^{-1}$ in atomic units, and this can be used to show that $\beta \propto t^{-1}$ if there is continuous creation of matter of the preferred ($\times$)-type (in which masses increase like t^2), or $\beta \propto t$ if there is no spontaneous mass creation. The relation $G \propto t^{-1}$ holds in both cases (since as shown by Canuto et al. (1977a) $G \propto \beta$ in the first case while $G \propto \beta^{-1}$ in the second case). Since the scale-covariant theory and the Dirac theory both use the LNH, most of the astrophysical consequences of the former are similar to those of the latter. However, a study of the astrophysical applications of the scale-covariant theory by Canuto *et al.* (1977c) has concentrated on showing how that theory might be connected up with elementary particle theory via astrophysical considerations, a topic which is of considerable importance (see Chapter 3) but which has not been examined in detail within the framework of the Dirac theory. Canuto *et al.* (1977c) have considered a scale-covariant cosmological model in which the strengths of the weak interaction of particle physics, the electromagnetic interaction and the gravitational interaction were all comparable in the early Universe (when the temperature was $T \approx 10^{15}$ K). Such a model is suggested by and is compatible with modern gauge theories of elementary particle physics. The different strengths of those interactions at the present epoch can be understood as the result of their gradual change with time, a process which can be accounted for in a natural way by the scale-covariant theory in terms of a time-dependence of the cosmological parameter Λ. (Other theories of similar types usually invoke a phase change in the early Universe to account for the present difference in strengths of the weak, electromagnetic and gravitational forces.) The scale-covariant theory involves a time-dependence of Λ according to $\Lambda \propto t^{-2}$, which implies that $\beta \propto t^{-1}$, since $\Lambda \propto \beta^2$ in this theory, and adds support to the idea of (×)-creation. It is interesting to note that $\Lambda \propto t^{-2}$ has also been deduced by Wesson (1978a) on the independent ground of wishing to find a cosmology that is scale-free in the sense of being self-similar (see Chapter 6) and yet incorporates a Λ term (the conventional Λ term defines a cosmological scale). Canuto *et al.* have also shown that the scale-covariant theory, when used in conjunction with a Robertson/Walker metric for a uniform Universe model, predicts a deceleration parameter of size $q_0 \approx 0$, which to within the large uncertainties attached to attempts at determining this parameter is compatible with observation.

The two main problems which the Dirac and Hoyle/Narlikar theories have been faced with, namely those to do with the past temperature of the Earth and the 3 K microwave background, have been studied in terms of the scale-covariant theory by Canuto and Hsieh (1978a, 1978b, respectively). As far as the temperature problem is concerned, the scale-covariant theory can lead to several different types of behaviour for $T_E(t)$, depending on whether G-variability is or is not accompanied by continuous creation, and on what form the latter process takes if it occurs. (The theory is compatible with the primitive, $G \propto t^{-1}$, no creation form of the LNH (Dirac, 1938), as well as with its more modern form (Dirac, 1973) in which continuous creation occurs in accordance with an increase in the number ($\approx 10^{80}$) of particles in the Universe as t^2.) Four special cases have been examined by Canuto and Hsieh

(1978a) for variations of G, $M_\odot$, the Earth/Sun distance (r) and the corresponding gauge function (β): (1) conventional theory (G, $M_\odot$, r and β all constants); (2) primitive Dirac (1938) LNH theory ($G \propto t^{-1}$, $M_\odot$ = constant, $r \propto t$, β = constant); (3) scale-covariant theory without matter creation ($G \propto t^{-1}$, $M_\odot$ = constant, $r \propto t^{-1}$, $\beta \propto t$); (4) scale-covariant theory with matter creation ($G \propto t^{-1}$, $M_\odot \propto t^2$, $r \propto t$, $\beta \propto t^{-1}$). They have concluded that case (3), which leads to a moderate decrease of T_E with time, is in best agreement with geological data which tend to show that the early temperature of the Earth was somewhat higher than it is now (Knauth and Epstein, 1976; see also Chapter 6). But since paleoclimatology is a rather inexact science, undue weight should not be attached to this conclusion, and case (4), which leads to a moderate increase of T_E with time, is also quite viable. The important thing is that, within the uncertainties inherent in the data, there are forms of the scale-covariant theory that do not conflict with geological evidence on the past temperature of the Earth.

As far as the problem of the 3 K microwave background is concerned, Canuto and Hsieh (1978b) have shown that it is possible to avoid the undesirable consequences of the Dirac theory (i.e., the implausible nature of the intergalactic medium and the expected departures from a black-body spectrum which characterizes Dirac's formulation of the LNH) by choosing a new gauge $\beta \propto t^{1/2}$. This is a departure from previous work on variable-G cosmology within the framework of the gauge geometry of Weyl, since hitherto attention has been focussed mainly on the two Dirac gauges $\beta \propto t$ ($G \propto t^{-1}$, (+)-creation, masses constant) and $\beta \propto t^{-1}$ ($G \propto t^{-1}$, (×)-creation, masses increasing as t^2). However, the new gauge is still consistent with the LNH to a certain degree since $G\beta^2$ = constant holds, so that $G \propto t^{-1}$ as usual. Canuto and Hsieh have demonstrated that the existence and spectrum of the 3 K microwave background are consistent with the LNH for a choice of gauge $\beta \propto t^{1/2}$, since it can be proved that if parameters are chosen suitably the radiation properties of the new model become indistinguishable from those of conventional cosmology. They have also shown that the 3 K field can be used to obtain information on the space-time curvature and scale factor of variable-G ($\beta \propto t^{1/2}$) cosmological models, one fairly definite inference being that space-time is flat (i.e., $k = 0$ for a Robertson/Walker metric) in such cosmologies.

The $\beta \propto t^{1/2}$ gauge thus helps the scale-covariant theory to avoid the problems encountered with the two Dirac gauges in regard to the 3 K microwave background. Further evidence in favour of the $\beta \propto t^{1/2}$ gauge comes from a study of Mach's Principle (Canuto et al., 1978). Although, as mentioned above, Mach's Principle is not precisely defined, many workers would regard with more favour a gauge that is compatible with such a principle over one that is not. In the scale-covariant theory, some choices of β lead to solutions that are Machian, while some do not. (A solution of the basic one-body problem for a gauge-covariant theory of the Dirac type has been given by Adams (1978) for the $\beta \propto t^{-1}$, ($\times$)-creation case.) Of the three previously discussed gauges, Canuto *et al.* (1978) have found that the $\beta \propto t$ and $\beta \propto t^{1/2}$ gauges are Machian while the $\beta \propto t^{-1}$, $\Lambda \neq 0$ gauge is not Machian. In view of the fact that the $\beta \propto t$ gauge is not in good agreement with observation, a belief in Mach's Principle therefore implies a preference for the $\beta \propto t^{1/2}$ gauge.

It can be seen from the outline given here that the scale-covariant theory of Canuto *et al.* is in agreement with observation as far as its consequences have been examined to date (Canuto *et al.*, 1977a, b, c; Canuto and Hsieh, 1978a, b). There remains a considerable scope for development of the theory, particularly in relation to non-stellar astrophysical systems (i.e., galaxies, clusters of galaxies and so on) and their evolution with (atomic) time. While at least one choice of gauge exists the consequences of which are in agreement with the 3 K microwave background, more astrophysical data of other types are needed before a final decision can be made about the relationship of the time scale of atomic processes to that of gravitational processes. But irrespective of this, the success to date of the theory of Canuto *et al.* makes it almost certain that scale covariance (or invariance) is a property that is compatible with the observed Universe.

2.3. The Status of Variable-G Gravitation

It is apparent from the previous section that the theories of Dirac, Hoyle/Narlikar and Canuto *et al.* have a lot in common: (i) They share the property of being able to make practical use of different time scales to illustrate and expedite the working out of astrophysical processes; (ii) they are all based on invariance under gauge or conformal transformations of the type given by (2.3.), and for mathematical purposes the terms gauge, conformal and scale invariance (or covariance) are equivalent as they are used in these theories (the term cocovariance is also used to describe scale-covariant theories of the Dirac type); (iii) the equation of geodesic motion, which describes how particles move through space-time, can be expressed in the same form for all three theories (Bouvier and Maeder, 1978). These things show that the theories (a), (b) and (c) have a considerable degree of technical overlap, although they are not completely equivalent.

The variation of G which characterizes all three theories is a process which occurs at a rate of order $|\dot{G}/G| \approx 10^{-11} - 10^{-10}$ yr^{-1}. The Large Numbers Hypothesis which forms the basis of the Dirac theory leads one to expect $\dot{G}/G \simeq -6 \times 10^{-11}$ yr^{-1} as noted previously. The theory of Canuto *et al.* predicts a comparable rate insofar as it makes use of the LNH, although that theory in general form is compatible with an arbitrary value of $\dot{G}/G$. The Hoyle/Narlikar theory (b) does not use the LNH either as a motivation as in (a) or as a gauge-fixing condition as in (c), and the rate of variation of G in this theory can be adjusted as desired by a choice of parameters (although the standard $G \propto t^{-1}$ choice gives $\dot{G}/G \simeq -1 \times 10^{-10}$ yr^{-1}). Data on a possible variation of G are available from astrophysics (Chapter 4), geophysics (Chapter 5) and gravity experiments (Chapter 6). The *reliability* of attempts at evaluating $\dot{G}/G$ increases with the field of research concerned in the order quoted, but the *exactitude* of the results decreases in the same order. Thus there exist limits on $\dot{G}/G$ derived by different methods with sizes of a few parts in 10^{12} yr^{-1} (astrophysics), 10^{11} yr^{-1} (geophysics) and 10^{10} yr^{-1} (gravity experiments). The more exact of these upper limits on $\dot{G}/G$ involve more ancillary assumptions than do the less exact ones. Thus, formally speaking, all three theories are viable because none of them contradict the limit $|\dot{G}/G| \lesssim 1 - 4 \times 10^{-10}$ yr^{-1} set by gravity experiments involving radar ranging of the planets. (See Chapter 6 for a discussion of ways of determining G-variability and a comparison of the results on $\dot{G}/G$ obtained from the different methods that have been employed to date.) Limits from different areas of physics that would be affected by a variation in G represent negative outcomes of attempts at measuring G-variability. The only result on the positive side is the finite determination by Van Flandern (1975) of $\dot{G}/G \simeq -8 \times 10^{-11}$ yr^{-1} which he arrived at by a study of the evolution of the Moon's orbit as revealed by lunar occultations of stars (see Chapters 5 and 6). Van Flandern's result can in principle be taken as support for any of the theories discussed above.

The accounts of gravitation represented by (a), (b) and (c) above are not the only viable theories of gravity available in modern physics. In particular, there are other metric theories of gravity which, like the three examined in this chapter, are compatible with the basic Solar System tests of relativity and so represent formally acceptable accounts of gravitation. A compendium of viable theories of gravity is given in Chapter 9. Some of the theories examined there involve G-variability, but in general they are not as well formulated or as extensively worked out as the theories of Dirac, Hoyle/Narlikar and Canuto *et al.* These latter represent the strongest contenders for a possible successor to Einstein's theory of general relativity, because they have strong theoretical foundations and are (more or less) in agreement with observation.

This is not to say that (a), (b) and (c) are equally 'good': each one of these theories has its virtues and its drawbacks. The Dirac theory has an attractive, clear-cut basis in the Large Numbers Hypothesis, but there is a possibility that it will founder because of problems in reconciling it with the 3 K microwave background. The Hoyle/Narlikar theory is not bound by the LNH and has the desirable attribute of being compatible with the absorber theory of radiation; but of the three main variable-G theories, it is in poorest agreement with astrophysical data and is in danger of losing its motivational basis because it is becoming increasingly likely that no redshifts exist in cosmology that cannot be explained as a manifestation of the Doppler effect in an expanding Universe (or an equivlent process if a static-Universe gauge is used). The theory of Canuto *et al.* only uses the LNH in a secondary role, but it is logically well-founded and in overall agreement with observation, although further development of the theory may reveal aspects in which it is in conflict with astrophysical data.

Overall, it would appear that the theories of Dirac and Canuto *et al.* are slightly favoured over that of Hoyle/Narlikar, but a lot of weight should not be attached to this comment because the balance of favourability can shift depending on developments in fields of research other than astrophysics and cosmology. Most notably, advances in particle physics may provide new pointers to an acceptable theory of gravity to supercede general relativity, and further developments in both areas are likely to come predominantly from the study of ways in which microphysics and macrophysics can be unified.

CHAPTER 3

PARTICLE PHYSICS AND GRAVITATION

3.1. Introduction

In the previous chapter it was seen how modern theories of gravity attach considerable importance to gauge (or scale) invariance. In the present chapter, attention will be directed towards gauge invariance in elementary particle theory and ways in which particle physics might be connected up with gravitation.

In quantum theory, the condition that the description of a particle system be independent of the gauge is ensured by the fact that the Schrödinger and Dirac equations (which describe non-relativistic and special-relativistic systems respectively) are invariant under gauge transformations (Yang, 1976a, b). Mathematically, this means that the formal structure of Hamiltonian quantum theory is unaffected by the presence of a gauge function $\chi(\mathbf{r}, t)$. Experimentally, it means that observable properties of a given system are the same irrespective of the choice of the gauge function.

Take, for example, a system characterized by electric and magnetic fields $\mathbf{E}(\mathbf{r}, t)$ and $\mathbf{B}(\mathbf{r}, t)$ derivable from potentials $\phi(\mathbf{r}, t)$ and $\mathbf{A}(\mathbf{r}, t)$, where

$$\mathbf{E} = -\nabla \phi - \frac{1}{c} \frac{\partial \mathbf{A}}{\partial t}$$

$$\mathbf{B} = \nabla \times \mathbf{A}.$$

(3.1)

Then the laws of classical particle physics are unchanged when a transformation is made to a new gauge in which

$$\mathbf{A}' = \mathbf{A} + \nabla \chi$$

$$\phi' = \phi - \frac{1}{c} \frac{\partial \chi}{\partial t}.$$

(3.2)

Under a gauge transformation of this type, the wave-function ψ for a particle system that is described by the wave equation with Hamiltonian H, $i\hbar(\partial \psi/\partial t) = H\psi$, changes from $\psi(\mathbf{r}, t)$ to $\psi'(\mathbf{r}, t)$ where

$$\psi' = \psi \exp{(ie\chi/c\hbar)}.$$

(3.3)

Here, e is the electron charge and $\hbar$ is Planck's constant h divided by 2π, as usual. The formal structure of the quantum theory of a particle in an electromagnetic field is unchanged by transformations of the type (3.2) and (3.3), showing that the theory is gauge invariant in this sense. (A transformation of the type (3.2) is known as a gauge transformation of the first kind, while in this context (3.3) is known as a gauge transformation of the second kind: see Rose (1961) and Chapters 8, 9 and 10.) The experimental success of non-relativistic and special-relativistic quantum theory, based on the Schrödinger and Dirac forms of the wave equation, shows that particle physics as it involves the electromagnetic interaction is fully consistent with gauge invariance in flat space-time (Yang, 1976a, b; Rose, 1961; Taylor *et al.*, 1969; Gupta, 1978; see Chapman and Leiter (1976) for a discussion of the covariant Dirac equation in the curved space-time of general relativity). This has led to the strong belief that modern theories of the weak and strong interactions, based on spontaneously-broken symmetry groups for descriptions of the particle parameters, should also be gauge invariant. It has led also to the wish to find a gauge invariant theory of gravitation.

3.2. Gauge Theories and Gravitation

Although there is, as yet, no firm connection between elementary particle theories with broken symmetries and gravitation, there is a widespread feeling that the partial success of the former has some connection with the latter, possibly via the cosmological constant Λ. A reasonably successful gauge theory of leptons was formulated by Weinberg (1967) and Salam (1968), and there is optimism that a gauge theory of elementary particles will be found which is in overall agreement with observation. There are many such theories (for reviews, see Abers and Lee, 1973; Fradkin and Tyutin, 1974; Weinberg, 1974a; Taylor, 1976). They are of interest because they hold out the hope of a realistic unification of the weak interaction (responsible for β-decay) and the electromagnetic interaction; and possibly of the strong (i.e., nuclear) and gravitational interactions as well (Weinberg, 1974a, 1977a, b, c; Bég, 1974; Freedman and Nieuwenhuizen, 1978). The Weinberg/Salam model shows how the weak and electromagnetic forces can be interpreted as the manifestations of a gauge theory based on a gauge group $SU(2) \otimes U(1)$ which is compounded of two smaller groups. Research aimed at finding an even larger group which can also account for gravitational forces has so far concentrated on hypotheses about the nature of the Λ term.

From the point of view of incorporating particle physics into a tensor theory of gravitation like Einstein's, the simplest approach is to interpret Λ in terms of quantum mechanics and the physics of the vacuum (Ford, 1975;

Steeruwitz, 1975; Wesson, 1978a). There have been three notable approaches to this problem:

(a) A straightforward comparison of general relativity with the Weinberg/Salam gauge theory allows one to calculate Λ in terms of particle properties, including the weak coupling constant (Dreitlein, 1974). A finite value of Λ implies a finite mass for the Higgs particles of the gauge field (according to Weinberg (1976a) a lower bound for the mass is 3.72 GeV/c^2). A positive value for the mass of the Higgs particles leads to a negative value for Λ, which in the cosmological context implies that the Universe will eventually cease expanding and recollapse to a ($\Lambda < 0$) singularity similar to that from which it began (Tipler, 1976; see also Ginzburg, 1971; Kharbediya, 1976; Ellis, 1978a). This is awkward, since cosmological data indicate that the expansion of the Universe is accelerating and that Λ is positive (Gunn and Tinsley, 1975). The evidence in favour of the latter conclusion has recently been re-examined (Tinsley, 1978a; see also Tinsley, 1977a): there is little doubt that data on the ages of the stars and the mean density of the Universe (as calculated from the observed galaxies) indicate that the deceleration parameter is $q_0 \lesssim -1$ and that $\Lambda > 0$.

This conclusion rests largely on the fact that globular clusters are believed to have ages of around 1.6×10^{10} yr, and there seems little reason to doubt this figure. (The absolute size of Λ in terms of H at the present epoch is inferred to be given by $\Lambda/3H_0^2 \simeq 1.1 - 1.6$, the mean density of the Universe in terms of the Einstein/de Sitter critical density needed to halt the expansion being $\Omega = 0.05-0.3$.) An 'age' for the Universe that agrees to within the uncertainties involved with that just quoted can be calculated from data on H_0, q_0, stellar ages and the dating of nuclei that have been processed by nucleosynthesis (Kazanas et al., 1978). The result is a self-consistent age of $1.35-1.55 \times 10^{10}$ yr (with $\Omega = 0.06-0.3$; an Ω in this range, besides being compatible with the noted data, is also compatible with the m/z diagram and the cosmological abundance of deuterium, these things indicating (Gott et al., 1974) that $\Omega \simeq 0.06$ (± 0.02), although values of Ω as high as $\Omega \simeq 0.1$ are acceptable.) Given the consensus of opinion on the age question, there are no grounds for doubting the conclusion that Λ as calculated from cosmological data is positive.

(b) Elementary particle theory with symmetry breaking can be connected with gravitational theory by making Λ dependent on the temperature T of the Universe (Linde, 1974). This is equivalent to interpreting the Lee constant of particle theory as a cosmological constant. However, the size of Λ in the early fireball stage of the Universe when T was high must then have been very large: $\Lambda c^{-2} > 10^{-6}$ cm^{-2} compared with the present value of $\Lambda c^{-2} <$

10^{-55} cm^{-2}. To reconcile this enormous discrepancy, the Universe is assumed to have passed through a phase transition not long after the big-bang event when $T \approx 10^{15}$ K (Linde, 1974; Kirzhnits and Linde, 1972, 1975, 1976). During and after the phase transition, Λ is supposed to have decreased in size as T decreased and the Universe expanded, most of the reduction being at early times when T was still near the phase-change value. The critical value of T at which the phase transition would have occurred (T_c) depends somewhat on the model considered; but very large values of T_c (like the one just quoted) are typical because T_c marks the transition between the regime ($T > T_c$) where the symmetry of the particle-interaction model is unbroken and the regime ($T < T_c$) where the symmetry is broken as it is in the present-day, low-temperature Universe (Weinberg, 1974b; see below). The phase-change argument in any particular form may or may not be plausible, but the general idea on which it is based has an important implication when re-expressed in another way (Wesson, 1978a): If elementary particle theory and gravitation have any connection with each other of the general type being considered in this chapter, then Λ must be non-zero, no matter how small its actual value may currently be. If this line of reasoning is followed, Λ cannot *a priori* be ignored in cosmological research.

(c) A way of avoiding the wrong sign of Λ given by approach (a) and the uncomfortable size of Λ given by approach (b), is to match broken-gauge particle theory not with a Universe having $T \equiv 0$ K as in (a) or with $T \approx 10^{15}$ K as in (b), but instead with a Universe at the microwave background temperature of $T \simeq 3$ K (Canuto and Lee, 1977; see also Canuto *et al.*, 1977c). This of course involves the assumption that the microwave background truly represents the thermodynamical ground state of the Universe.

The microwave background is believed by almost all astrophysicists to be the cooled-down radiation from the big-bang fireball (see Chapter 11). However, there are still occasional claims that the background is not the remnant of a primeval fireball (see Peebles (1971a) for a critical discussion), and that it might be the result of non-cosmological processes acting at the present epoch, such as radiation from cosmic dust, for example (Wickramasinghe *et al.*, 1975). These claims are suspect because they do not account for the observed isotropy of the radiation (Wesson and Lermann, 1976; Chapter 11). Alternatively, it has been suggested by Alfvén and Mendis (1977) that the microwave background represents radiation that was thermalized by dust in galaxies at an early epoch in the history of the Universe, when the latter was opaque to the background field (this epoch corresponding to a redshift of $z \simeq 40$). The opacity calculation depends, however, on grain properties, and

Pollaine (1978) has shown that the optical depth of the Universe was probably not larger than 0.1 back to the decoupling epoch (corresponding to $z = 1500$). Thus, the accurately black-body form of the microwave spectrum implies that thermalization occurred at quite large redshifts. Carr and Rees (1977) pointed out that the background might have been generated in the early Universe when any primordial black holes that were present would have been accreting matter from their surroundings and emitting energy. This process can produce enough photons to account for the observed photon/baryon ratio (10^8–10^9) by which microwave photons outnumber protons at the present time, but again it is not clear that it explains the thermal form of the background spectrum. (One way of obtaining a thermal spectrum is to identify the microwave background photons with particles produced by evaporating singularities either of the black hole (Hawking, 1974, 1975, 1976a, b, 1977; Davies and Taylor, 1974; Davies, 1976a, b, 1977a; Kundt, 1976; Gibbons, 1976; Carr, 1976) or big-bang (Hawking, 1976b; Parker, 1976; Davies, 1977a) varieties.) A related proposal by Rees (1978a) is that the background was produced in the pre-galactic era of a big-bang cosmology by a population of massive stars that burnt up about 0.5% of their rest mass energy, emitting this as radiation which was then thermalized by gas and dust. (This idea was due in original form to Layzer (1968a, b) and Layzer and Hively (1973), but in their version of it the energy production was supposed to take place at uncomfortably small redshifts of $25 \lesssim z \lesssim 50$.) The hypothesis in the form given by Rees also accounts for the value of the photon/baryon ratio in the Universe, and is not as open to criticism as other hypotheses of a similar type because the generation and thermalization would have occurred very early on, probably at epochs $t \lesssim 3 \times 10^6$ yr or at redshifts $z \gtrsim 200$. Thermalization may even have occurred at $z \gtrsim 10^4$ as discussed by Rees (1972a; see Longair (1976), Danese and de Zotti (1977, 1978) and Weinberg (1977d) for reviews of the relic radiation spectrum and the thermal history of the Universe). Accepting, therefore, the conventional view that the microwave background is of early ($z \gg 1$) big-bang origin and all-pervasive, it is certainly natural to compare gauge theories of elementary particles with a Universe having $T \simeq 3$ K.

An analysis using the Weinberg/Salam theory, the Mohapatra/Sidhu theory and the σ-model of elementary particles reveals that the last can most easily be brought into conformity with astrophysical data (Canuto and Lee, 1977). The Λ involved is positive in sign and small in size.

While it may not yet be clear how to link up gauge theories of the elementary particles and gravitation, optimism exists about the possibility of finding a gauge-invariant theory of gravity. The main ground for this optimism is that the gauge theories of elementary particles (in one or another form) are in reasonable agreement with experiment, insofar as one considers theories like that of Weinberg and Salam which unite the weak and electromagnetic forces (Leader and Williams, 1975; Wilkinson, 1975; Blin-Stoyle, 1975). While the methods of particle theory can also be employed to bring the strong interaction into this scheme, it must nevertheless be admitted that there is a conceptual problem involved when one attempts to bring in gravity as well. This problem is that it is not easy to match the formalism of particle theory with that of a metric theory of gravity such as general relativity. Mathematically, it is not clear how one should construct a Lagrangian that includes both particle physics and gravitation. Equivalently, from the other side, it is not clear what energy-momentum tensor one should put on the right-hand side of Einstein's field equations in the case where one wishes to obtain a geometric theory of a space-time in which the non-gravitational interactions of particles are included. From the latter point of view, one can say that the main thing which is lacking for a unified theory of gravitation and particle physics is an energy-momentum tensor for the vacuum. (In Einstein's theory, the term in the cosmological constant Λ represents the properties of the vacuum via the equation of state $p_\Lambda = -\rho_\Lambda c^2$ where $\rho_\Lambda = \Lambda/8\pi G$, but this is only a gross way of describing the vacuum (Wesson, 1978a). One would really like to replace Λ by a term that gives a proper description of the physics of the vacuum, as T_{ij} does for the matter.) The physics of pair creation and similar processes do, of course, give some information about the vacuum; but in other regions of physics, and in particular in cases where gravity is known to be dominant as in the astrophysical domain, very little is known about the properties of the vacuum.

However, it is not yet clear if vacuum processes have any significant implications on the cosmological scale. For example, the Davies/Unruh effect (Davies, 1975, 1976b; Unruh, 1976) leads one to expect that an observer who accelerates with respect to the random, zero-point fluctuations of the electromagnetic field of the vacuum should detect a thermal background around himself, the temperature of which would be related to the value of his acceleration. (This effect has been reviewed by Sciama (1978) who has also discussed its astrophysical implications.) But to obtain even a temperature of $T \simeq 3$ K as observed in the microwave background would require an acceleration of 10^{20} cm sec^{-2}. This is enormous even by astronomical standards, and it is difficult to see what relevance vacuum processes such as this have for astrophysics. This is not to say, though, that the vacuum part of space-time is totally irrelevant to gravitational phenomena in astronomical systems. Limits on the size of Λ leave open the possibility that the vacuum density $\rho_\Lambda = \Lambda/8\pi G$ is much greater than the matter density ρ seen in the galaxies (Wesson, 1978a). This refers to the present epoch, and in view of the comments made above about a possible time-dependence of Λ, the Universe may have been even more Λ-dominated in the past than it is now. If the metric of space-time is dominated by the vacuum rather than (as is often tacitly assumed) by the matter in it, one might be able to explain a puzzling property of the Universe: namely, the absence of any anisotropy in the property of inertia due to the anisotropic distribution of matter on the astronomical scale (e.g., the anisotropy of the matter distribution of the Milky Way as seen from the Sun's off-centre location in the Galaxy). Wesson has discussed the origin and distance-dependency of the resistance to acceleration of matter which is known as inertia, and which one might reasonably expect to be due to an interaction with other matter in the Universe (Wesson, 1978a). It is well known that the inertial mass of a local body shows no anisotropy down to very low limits of at least 1 part in 10^{20}, and probably down to 5 parts in 10^{23} (Hughes and Williams, 1969). This means that space itself is very isotropic, even if the matter in the galaxies is not.

The preceding comments show that Λ, far from being a regrettable complication in Einstein's field equations, may prove to be one of the most crucial factors in physics. Its key role is that it provides a connection between gauge theories of elementary particles and gravitation; and although the nature of this connection is obscure, it is still possible to make estimates of the major consequences of extended gauge theories as they affect large-scale astrophysical systems. The remainder of this chapter will therefore be devoted to discussing the main implications of gauge-invariant particle theories for cosmology.

3.3. Particle Physics and Cosmology

A renormalized theory of the weak interactions of the Weinberg/Salam type is typified by a long-range character for the weak interaction in the early stages of hot big-bang models of the Universe (Kirzhnits and Linde, 1972). The effects of this long-range force at times when $T \approx 10^{15} - 10^{16}$ K and $\rho \approx 10^{29}$ gm cm^{-3} (i.e., at $t \approx 10^{-12}$ sec) are analagous to those of electromagnetism in models where a condition of electric neutrality is not imposed. In order to avoid enormous forces arising from the long-range nature of the weak interaction at early epochs, one has to impose a condition of neutrality in the 'weak charge' sense, as discussed by Kirzhnits (1972). As noted above, unified theories of the weak and electromagnetic interactions are characterized by spontaneously-broken symmetries which may be connected with a phase transition which takes place at some critical temperature T_c. Before the transition, weak and electromagnetic interactions (and perhaps strong interactions also) are of comparable strengths, only becoming disparate after the transition. Thus, if one identifies the critical temperature T_c as being a property of a big-bang fireball, one can, in principle, say that the symmetry breaking of the Weinberg/Salam type of theory is due to the cooling that takes place as the cosmological model expands from its origin in a singularity.

This approach, which was discussed above in connection with Λ, has been used by Domokos *et al.* (1975) to calculate that the weak interaction constant must be changing at a rate of less than 10^{-10} yr^{-1} at the present epoch, provided one identifies the combination of the Weinberg/Salam particle model and the Robertson/Walker cosmological model with the real Universe. Domokos *et al.* (1975) have calculated a value of $T_c \approx 10^5$ K for the critical temperature; but Davies (1976c) has objected to such a low value of T_c on the grounds that it would upset nucleosynthesis processes in the early stages of the model (e.g., the synthesis of helium which begins when T is about 10^9 K), and lead to a Universe composed entirely of hydrogen. The value of T_c depends on various things, not the least important of which is the nature of the cosmological model adopted and the way in which its matter properties depend on time (Domokos *et al.*, 1976). In view of problems like the one just mentioned which follows from taking a too-low value of T_c, it would seem better to adopt a value of $T_c \approx 10^{15}$ K as noted previously if one wishes to use the phase-transition model of the early Universe (Kirzhnits and Linde, 1972, 1975, 1976; Linde, 1974). However, irrespective of the precise value of T_c, if a phase transition really did occur in the early stages of our Universe, the qualitative aspects of that event must have had considerable effects on how matter in the fireball behaved. In particular, a phase transition would have notable implications for theories of galaxy formation.

An examination of some cosmological consequences of a high-temperature hadron phase transition by Bugrii and Trushevskii (1977) shows that at early epochs the strong (as opposed to weak) interaction could have been important in accounting for the properties of systems that later developed into galaxies. In their model, there is a period after $t \simeq 1 \times 10^{-12}$ sec during which matter is spontaneously produced everywhere and the density ρ is a constant even though the scale factor is growing as $S(t) \propto e^t$. This period is succeeded at $t \simeq 1 \times 10^{-9}$ sec by a condensation (phase transition) which is supposed to have formed the systems which are now observed as galaxies and inhomogeneities of other types in a Universe in which gravitation has become the dominant force. In a wider context, it is interesting to note that a recent form of the equation of state of matter at high densities predicts the occurrence of a phase transition (Canuto, 1974, 1975, 1977; Datta, 1977; Canuto et al., 1978). Used in the big-bang model, the occurrence of such a phase transition predicts the formation of mini black holes which might have nucleated galaxy formation (Canuto, 1978). While the equation of state used by Canuto is typical of matter at neutron-star densities, a phase transition is also possible at lower densities of order 1 gm cm^{-3} in hydrogen at zero temperature (Zeldovich, 1963). This has been used as the basis of an account of the formation of condensations in the early Universe by Layzer (1969) and Hively (1971), but has not been connected with gauge theories of elementary particles. It can be noted, though, that the likelihood of phase transitions having occurred in the big-bang fireball (whether hot or cold) gains collateral support from the fact that only those theories which involve a phase transition give a ready account of why galaxies exist today. (Other work on this topic is discussed in Chapter 11; but an adequate theory of how galaxies form, come to have angular momentum, evolve into their different types, and come to be clumped together into clusters and superclusters, is still lacking.) Further research on these matters may show that the disparate strengths of the forces of physics and the existence of the clumpy structure of the Universe are both consequences of a phase transition that occurred in the primeval fireball.

Another way in which gauge-invariant quantum theories may affect astrophysics concerns the possible existence of new types of particles. In particular, some gauge theories of the weak interactions predict the existence of neutral heavy leptons, and these can have cosmological consequences of greater or lesser significance depending on the mass of the particle concerned. By considering the equilibrium thermodynamics of such particles in the early stages of the big bang when the temperature was $T \approx 10^{10}$ K, it is possible to infer a lower limit for the mass of the hypothetical neutral heavy lepton of $2 \text{GeV}/c^2$ if that particle is stable (Lee and Weinberg, 1977). This limit is arrived at by a comparison of the expected mass density, due to survivals from lepton/antilepton annihilation in the fireball, with the dynamically-inferred mass content of the presently-observed Universe. (A related study using a unified gauge theory of particle interactions has been made by Yoshimura (1978) and accounts for the dominance of protons over antiprotons in the present Universe and the entropy of the 3 K background; this approach is discussed further in sub-section (5d) of Chapter 9.) If neutral heavy leptons are not stable but liable to decay, then astrophysical observations can again be used, this time to obtain an estimate for their mass of 10 MeV/c^2 (Dicus et al., 1977). This estimate is arrived at by a comparison of the expected neutrino decay products of neutral leptons with constraints on a possible cosmological neutrino background in the observed Universe.

The difference of about a factor 100 between the two noted estimates for the mass of the hypothetical neutral heavy lepton means that such leptons can have a significant cosmological population either in the form of relatively heavy particles or in the form of relatively light particles. In principle, the two quoted studies leave open the possibility that neutral heavy leptons can have serious effects on astrophysics and cosmology, such as providing enough mass – in addition to that seen in the galaxies – to gravitationally bind the Universe and lead to a halting of its expansion with a subsequent recollapse to a singularity. (This possibility has been discussed by Davies (1977b), while Close (1978) and Perl (1978) have discussed recent progress in heavy lepton physics.) The situation is not, however, clear-cut since it depends on how the heavy leptons interacted with other matter in the early history of the Universe; and a larger mass for the heavy lepton does not necessarily mean that such leptons will contribute more to the cosmological mass density of the Universe at the present epoch. It has been calculated by Gunn *et al.* (1978) that if the mass of the heavy, stable neutral lepton is of the order of a few GeV/c^2, as seems likely, hidden mass in the form of such leptons could provide the dark matter believed to be present in galactic haloes. Such leptons might also supply the 'missing' mass in clusters of galaxies, although they cannot represent enough matter to gravitationally bind the Universe and lead to the halting of its expansion if they have masses of the order noted. Since the existence of neutral leptons of the type being considered can be accommodated into cosmological models without seriously upsetting big-bang nucleosynthesis calculations, confirmation of their presence would open up the possibility of solving some long-standing problems in astrophysics without departing too much from the framework of conventional cosmology.

CHAPTER 4

ASTROPHYSICS

The purpose of this and the following chapter is to give short accounts of how a variation in G can manifest itself in astrophysics and geophysics respectively, with special reference to limits which can be set on $\dot{G}/G$ from these two disciplines.

Astrophysical systems are affected by a decrease of G in numerous ways, some of which have already been discussed in the two preceding chapters. The main consequences of a simple decrease in G as measured on atomic time are the following: (1) the distance from the Earth to the Sun changes; (2) the Sun and other stars become less luminous; (3) likewise, the Milky Way and galaxies in general become less luminous due mainly to the dimming of their component stars; (4) the orbits of stars around the centre of the Galaxy become less tightly bound to the nucleus, and the same is true of stellar orbits in galaxies in general; (5) the dynamics of clusters of stars and clusters of galaxies (on an atomic time scale) are altered compared to ordinary Newtonian theory such that the clusters become less strongly bound; (6) cosmological parameters such as those describing the cosmochemical history and dynamical future of the Universe are altered with respect to those calculated on the basis of conventional (G = constant) theory.

If continuous creation of matter operates in conjunction with G-variability, some of the above processes are modified. In particular, the Earth and the other planets can now either recede from or approach the Sun depending on what type of creation is involved and what theory is used to describe the process (see Chapter 2; in the Dirac theory, ($\times$)-creation causes a recession, while (+)-creation causes an approach). Also, the luminosity of the Sun and the galaxies can increase with time if creation of the Dirac ($\times$)-type occurs and astronomical bodies increase in mass as t^2 (on atomic time). As noted elsewhere, the effects of multiplicative creation are often such as to offset the consequences of G-variability, whereas the effects of additive creation leave the consequences of G-variability dominant.

Stellar luminosities, calculated as a function of time with G-variability and continuous creation taken into account, have been used to judge the acceptability of the Dirac theory as discussed in Chapter 2. It was also noted there that the only positive evidence in favour of a variation in G is the result of Van Flandern (1975) that $\dot{G}/G \simeq -8 \times 10^{-11}$ yr^{-1}. VandenBerg (1976,

1977) has examined in detail the effects of a variation in G at this rate on the luminosities of stars. He has found that for G-variability alone, the luminosity of a typical star when it turns off the main sequence would be 0.5–1.0 magnitudes fainter than that given by conventional theory (G = constant) and as observed in the stars which comprise globular clusters. Further, stars which evolve with a decreasing value of G would have luminosity functions which are in disagreement with those observed for actual stars in the Galaxy. These two disagreements of the G-variable hypothesis with observation imply that G-variability at the Van Flandern rate is unacceptable by itself (VandenBerg, 1976). However, when a $G \propto t^{-1}$ variation is combined with a variation of masses in proportion to t^2, stellar evolution calculations show that stars evolve in almost the same manner as in conventional (G = constant, masses = constant) theory. In this case, there is no disagreement with data from stars in globular clusters and old galactic clusters (VandenBerg, 1977). These results confirm the calculations which have been made to test the Dirac theory (Chapter 2), and show that as far as data on the brightnesses and colours of stars are concerned, G-variability is only acceptable when combined with a variation of masses.

This conclusion is also supported by data on the past temperature of the Earth and other bodies in the Solar System (see Chapters 2 and 6). On the basis of the primitive version of the Large Numbers Hypothesis, in which G decreases without any continuous creation of matter (Dirac, 1938), the Sun's luminosity would have been higher in the distant past. This would have meant a warmer environment for meteorites in particular in the early history of the Solar System, and Peebles and Dicke (1962a) have used this to put a limit on simple G-variability. The higher temperature would have caused an anomalous loss of argon from meteorites, and a comparison of the expected size of this effect with the ages of meteorites as obtained from the potassium-argon dating method leads to a limit of $|\dot{G}/G| \lesssim 1 \times 10^{-10}$ yr^{-1}. This limit is not as accurate as limits obtained from stellar evolution calculations, but it can be improved upon by making a reasonable assumption that links G-variability with elementary particle theory via the fine-structure constant $\alpha (\equiv e^2/\hbar c$ in cgs units or $e^2/4\pi\epsilon_0 \hbar c$ in mks units where e is the electron charge and ϵ_0 is the permittivity of free space). In this way (Peebles and Dicke, 1962b), data on meteorites and Earth rocks can be used to yield the limit $|\dot{G}/G| \lesssim 2 \times 10^{-11}$ yr^{-1} for simple G-variability.

The consequences of a simple variation in G are not only unacceptable as regards the luminosities of stars but are also unacceptable as regards the luminosities of galaxies composed mainly of stars (Chapter 2; Wesson, 1978a). Likewise, G-variability coupled with continuous creation of the multiplicative type is acceptable not only on the basis of stellar luminosities but also on the basis of galaxy luminosities. However, G-variability without continuous creation of matter can be made consistent with luminosity data provided one takes $|\dot{G}/G|$ small enough, so it is of interest to see what limits can be set on G-variability from other astrophysical data.

White dwarfs provide one way of investigating the effects of G-variability. Vila (1976) has found that if G decreases, there should exist maximum possible ages and minimum possible luminosities for white dwarfs of given masses that can be observed today (other white dwarfs which may have existed in the past having been destroyed in supernovae explosions). A comparison of these expected limits with observed white dwarfs shows that G-variability at about the Van Flandern rate (or less) is acceptable on the basis of these data. White dwarfs can also be used to investigate possible departures from Newton's law of gravitation of the type expected from some variable-G theories, such departures being expected to show up not in the luminosities but in the dynamical characteristics of white dwarfs pursuing orbits in the Galaxy and in star clusters (Finzi, 1962; 1963a, b; see also Chapter 9). However, no useful limits on G-variability have been obtained as yet from this method so far as it concerns white dwarfs.

Star clusters in general, though, have been used to gain information on G-variability by comparing computer simulations of the dynamics of stars in a cluster affected by a decreasing G with those of stars in a cluster with constant G. Angeletti and Giannone (1978) have carried out such a comparison, taking into account various non-cosmological effects that are significant for cluster evolution such as mass loss by the stars in the cluster due to astrophysical processes. These workers note that if one adopts a rate of change of G of about the size found by Van Flandern, and assumes that $\dot{G}/G$ = constant in time, then $G \propto e^{-t}$ and the dynamical evolution of a cluster of stars is seriously affected by such a variation because its influence is cumulative over the long history of such a system. Compared to a G = constant star cluster, Angeletti and Gianonne have found that a decrease in G tends to prevent formation of a dense core (a phenomenon characteristic of G = constant clusters) and leads to a general expansion, although there is no increased tendency for the cluster to lose its member stars. On the basis of their simulations and observations, they conclude that G-variability at about the Van Flandern rate (or less) is acceptable.

The reliability of this conclusion is not fully established. It has been pointed out by Marchant and Mansfield (1977) that studies of the evolution of N-body systems such as star clusters and galaxies cannot be used to obtain useful limits on $\dot{G}/G$ if the natural dynamical evolution time scale for the system (t_d)

is less than the time scale characteristic of the gravitational effects of a weakening gravity (t_g). Some astrophysical systems do have $t_d < t_g$ and in the reasonable case where $t_d \ll t_g$ and there is virial equilibrium, Marchant and Mansfield have shown that the qualitative evolution of Newtonian N-body systems will not reveal a variation of G. This does not, however, mean that such studies are a waste of time, as there are two cases in which simulation of N-body systems might yield information on G-variability, namely those in which two-body relaxation is important (this may apply to the work of Angeletti and Giannone, 1978) and those in which the condition $t_d \ll t_g$ does not hold. The latter possibility is particularly interesting and applies to the work of Lewis (1976) on the alleviation of the missing mass problem in groups of galaxies on the basis of G-variability (see Chapters 2 and 9; the cosmologies discussed there in which the usual form of the virial theorem does not apply are also exempt from the criticism of Marchant and Mansfield). Thus, provided one chooses systems for study in which the virial condition $t_d \ll t_g$ does not hold, and provided one keeps in mind that changes in the structure of gravitational systems due to G-variability can only be measured in practice against atomic time (or some process such as nucleosynthesis that effectively measures atomic time), one should be able to obtain useful information on G-variability from simulations of N-body systems such as clusters of stars and clusters of galaxies.

Cosmology can also be used to investigate the possibility of a variation of G, and in fact a study of the nucleosynthesis of elements in the big-bang origin of the Universe provides the best indirect limit on $\dot{G}/G$ that is available. Barrow (1978a) has examined the cosmochemical evolution of a standard big-bang model Universe as it might be affected by G-variability. The main data which he employs are the cosmic abundances of He^4 (which is $25^{+7}_{-3}\%$) and D (which is of order $10^{-3}\%$). The conventional big-bang model can account for these abundances in terms of high-temperature nuclear reactions in an expanding fireball, and a comparison of the conventional model with one which is affected by a variation of G leads to the very low limit of $|\dot{G}/G| \lesssim 1.5 \,(\pm 0.7) \times 10^{-12}$ yr^{-1} for the allowed rate of change of G at the present epoch. This limit depends, of course, on the validity of the standard big-bang model, but is nevertheless an impressive result which can only be reconciled with variable-G theories either by going outside the framework of accepted cosmology or by invoking variations of physical parameters other than G.

It was in fact suggested by Gamow (1967a) that G might be constant but that the electron charge e might change with time according to $e^2 \propto t$. Gamow made this proposition in order to keep the spirit of the Large Numbers

Hypothesis of Dirac but avoid the large luminosity change of the Sun and the corresponding large change in the temperature of the Earth over geological time which the primitive form of the LNH implies. (Gamow adopted the LNH in the form $e^2/4\pi\epsilon_0 G m_p^2 \approx 4\pi\epsilon_0 m_e c^3 t_0/e^2$, but actually this is numerically incorrect: the left-hand side is about 1×10^{36}; whereas the right-hand side, which is the age of the Universe $t_0 \simeq 10^{10}$ yr $\simeq 3 \times 10^{17}$ sec in terms of the unit of time $e^2/4\pi\epsilon_0 m_e c^3 \simeq 1 \times 10^{-23}$ sec, is about 3×10^{40}. Nevertheless, Gamow's hypothesis has been much discussed.) To be consistent, if G = constant and $e^2 \propto t$, then the electron mass m_e should also vary as $m_e \propto t$ (Gamow, 1967b). With this reformulation of the LNH, the relation which gives the dependency of the Sun's luminosity on the parameters involved (Gamow, 1967c) could be used to show that the temperature of the Earth's surface T_E might be expected to vary as $T_E \propto t^{-3/4}$ (Gamow, 1967d). This kind of dependency implies that the oceans would have been near to boiling 3×10^9 yr ago, a consequence which is a little uncomfortable (see Chapters 2 and 6) but might just be compatible with geological data.

Gamow's reformulation of the Large Numbers Hypothesis (G = constant, $e^2 \propto t$, $m_e \propto t$) succeeding in avoiding the strong dependency of T_E on time which is connected with that hypothesis in its primitive form ($G \propto t^{-1}$, e^2 = constant, m_e = constant), and it also provides a way of reconciling the LNH with the low limits which have been set more recently on a possible time-variation of G. It is doubly unfortunate, therefore, that Gamow's idea has turned out to be wrong.

Both the $e^2 \propto t$ and $m_e \propto t$ dependencies of Gamow's hypothesis have been effectively disproved by experiment and observation. As far as a possible variability of the electron charge is concerned, limits on a time-dependence of e began to appear soon after Gamow made his proposal. (For example, Gold (1968) set the limit $|\delta e/e| \lesssim 2.3 \times 10^{-4}$ for a change in e over the last 2×10^9 yr from a study of fission decay in rocks.) A review of limits on possible variations of the parameters of physics, including e, has been given by Wesson (1978a). Numerous data exist which preclude any variation of e, the best limit (Dyson, 1967) being $(e^2)\dot{}/e^2 \lesssim 3 \times 10^{-13}$ yr^{-1}. As far as possible variability of the electron mass is concerned, there is collateral proof against any variability on time scales less than of order 10^{11} yr, and there is direct proof from two sources that the electron-proton mass ratio (m_e/m_p) does not vary significantly with time: Firstly, Yahil (1974) has shown that m_e/m_p cannot have varied at a rate of more than 1.2×10^{-10} yr^{-1} over the last 10^9–10^{10} yr on the basis of the concordance of potassium-argon and rubidium-strontium age determinations. Secondly, Pagel (1977) has set the limit $[\ln (m_e/m_p)]\dot{} \lesssim 5 \times 10^{-11}$ yr^{-1} (with a similar limit for the gyromagnetic ratio of the proton) from a study of quasar absorption line redshifts. Thus, one sees that neither e nor m_e vary significantly on a cosmological time scale ($\approx 10^{10}$ yr). This means that one cannot invoke a variability of other parameters in order to circumvent astrophysical and cosmological limits on G-variability while retaining the Large Numbers Hypothesis.

Besides limits on a possible time-dependence of G, e and m_e/m_p, there are also limits on a range of other parameters, including the weak and strong coupling constants, the fine structure constant (α) and Planck's constant (h). Limits on $\dot\alpha/\alpha$ stand at a few parts in 10^{11} yr^{-1} by reasonably direct determinations and at a few parts in 10^{17} yr^{-1} by less direct determinations. Limits on $\dot h/h (= \dot\hbar/\hbar)$ stand at a few parts in 10^{14} yr^{-1} (Wesson, 1978a; see also Chapter 2). Limits on a possible variation of $\hbar$ are of particular interest since $\hbar$ is the most fundamental parameter of elementary particle theory. Wasserman and Brecher (1978) have reconfirmed that the value of $\hbar$ was not significantly different from the present value in the early Universe by a calculation similar to that of Barrow outlined above. Wasserman and Brecher have studied the nucleosynthesis of D and He^4 in the standard big-bang model of the Universe, and have used data on the observed abundances of these isotopes in stars today, to deduce that the value of $\hbar$ in the first few seconds of the fireball was within a factor three of its value at the present epoch. To be precise, if $\hbar_{nuc}$ was the value of $\hbar$ at the epoch of nucleosynthesis ($t = 10^{-1} - 10^3$ sec, corresponding to redshifts of $z = 10^8 - 10^{10}$), and $\hbar_{now}$ is its value now, then Wasserman and Brecher have shown that $0.2 \lesssim \hbar_{nuc}/\hbar_{now} \lesssim 0.3$ holds. Limits like this and the ones noted above and elsewhere on possible variations of the parameters of physics other than G can be summed up by saying that the constants of physics are indeed true constants, at least on time scales of the order of the age of the Universe ($\approx 10^{10}$ yr) and probably on time scales of 10^4 times as long.

This puts G-variability in a somewhat awkward situation. Although many modern theories of gravity are compatible with a time-variation of G, whereas modern theories of elementary particles are mostly not compatible with a time variation of the nuclear and atomic parameters, one feels that there ought nevertheless to be some connection between the two disciplines. If so, it seems unreasonable to expect that G should vary but that the other parameters of physics should not. The various arguments that have been given for believing that the gravitational interaction should be time-variable whereas the other interactions should be time-independent (Wesson, 1978a) boil down in the last analysis to the fact that gravity is a weak force whose nature is probably linked fundamentally to the mass properties of a Universe which is itself evolving and expanding. This argument may be plausible, but it is likely that simple G-variability is a naïve interpretation of it. The overall success of the Dirac ($\times$)-model as regards astrophysics implies that if G varies then masses probably vary also. It is quite likely that a complete theory of gravitation which is in agreement with all astrophysical data will prove to be one in which several cosmological processes of this type operate together, with G-variability being only one of a series of interdependent effects.

CHAPTER 5

GEOPHYSICS

The Earth, by virtue of its rotation, represents a kind of cosmic clock that has been running for about 4.5×10^9 yr. During that time, it has been gradually slowing down, at a rate of about 2 msec a century, and by estimating the various processes that contribute to this spin-down it is possible in theory to gain information on long-term cosmological effects such as G-variability and continuous creation. In practice, this is a complicated problem because there are various purely geophysical things which contribute to the spin-down. Most of the deceleration is believed to be due to tides in the oceans which are raised on the Earth by the gravitational attraction of the Moon, and which dissipate energy, and so lead to a decrease in the angular velocity of rotation. The energy so lost by the Earth is transferred to the Moon, which therefore recedes, and measurements of the acceleration of the Moon in its orbit provide one method of estimating the spin-down rate of the Earth. Cosmological effects, such as a decrease in G, continuous creation and expansion of the Earth (possibly caused by G-variability or matter creation) must be evaluated in conjunction with the tidal term and other geophysical processes that can alter the planet's rate of spin.

The use of the rotation of the Earth as a means of estimating cosmological effects such as those just mentioned has been reviewed by Wesson (1978a), and related measurements are discussed below in Chapter 6. Data which involve the Moon as a way of estimating the rotational history of the Earth are available from lunar occultations of stars monitored directly against atomic time (for the last few decades), telescope observations (for the last few centuries), records of eclipses (for the last few thousand years) and records of growth lines in some types of fossil which reflect the ancient sequence of the days and months (for the last few 10^8 yr). The interpretation and intercomparison of these sets of data is not completely free of ambiguity, but the majority opinion is that they are not compatible with a simple G-variability (without creation) at a rate larger than $|\dot{G}/G| \simeq 4 \times 10^{-11}$ yr^{-1}, and probably restrict G-variability to $|\dot{G}/G| \lesssim 2 \times 10^{-11}$ yr^{-1}. One exception is the result of Van Flandern (1975) which is based on lunar occultation data and indicates G-variability at a rate of $\dot{G}/G \simeq -8 \times 10^{-11}$ yr^{-1}. (The exactitude of this determination is difficult to assess, the mean error being about $\pm 3 \times 10^{-11}$ yr^{-1} as far as random sources of error are concerned or $\pm 5 \times 10^{-11}$ yr^{-1}

if one takes into account possible systematic sources of error. The result as it stands is probably significant.) If this result is accepted, one has to invoke compensating processes in order to reconcile such a variation of G with other data on the dynamical behaviour of the Earth-Moon system. This is only to be expected on theoretical grounds, and the past rotation of the Earth as inferred from non-lunar data is consistent with G-variability at about the rate found by Van Flandern (see Chapter 6). Processes which tend to affect G-variability, such as continuous creation or the reorganization of the material of the Earth's interior, leave considerable latitude for the presence of a G-variable term in the equations which govern the long-term behaviour of the Earth-Moon system. For example, Dirac's ($\times$)-model, which involves creation and G-variability at a rate of $\dot{G}/G \simeq -6 \times 10^{-11}$ yr^{-1}, is compatible with data on the rotation of the Earth.

Expansion is a complementary process to G-variability and continuous creation of matter, and most of this chapter will be devoted to a short examination of the evidence for a cosmologically-caused expansion of the Earth. In Chapter 2 it was seen that the Hoyle/Narlikar theory predicts an expansion rate of about 0.1 mm yr^{-1}; while the scale-covariant theory, using Dirac's Large Numbers Hypothesis as a basis, predicts a rate of 0.02–0.03 mm yr^{-1} if there is no matter creation and 0.2–0.3 mm yr^{-1} if there is matter creation. The question therefore is whether geophysical evidence indicates expansion, and if so at what rate.

The expanding Earth hypothesis has had a long and uneven history. Reviews of its status have been given by Dearnley (1966), Creer (1967), Carey (1976) and Wesson (1973, 1978a). A considerable body of data on estimates of the past radius of the Earth has been collected and discussed by Creer (1967), and those data are plotted in Figure 1 as estimates of the rate of expansion over geological time. Ways of estimating the past radius can conveniently be divided up into those that do not make use of palaeomagnetism and those that do, and the data of Figure 1 have been plotted according to this grouping (see below for a comparison of the reliability of the results obtained by the two methods). Since Creer's compilation, there has been sporadic interest in the expanding Earth hypothesis, but until recently there was no consensus of opinion about what might be causing the expansion (if it exists). Modern variable-G cosmologies provide a plausible basis for expecting that the Earth should be expanding, but this is not to say that geophysicists are at all in agreement either as regards the rate of expansion or indeed as regards its existence.

If expansion is occurring, it is unlikely to be doing so at the relatively fast rate of 6–7 mm yr^{-1} suggested by Carey on the basis of geological data

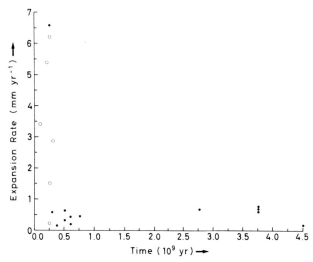

Fig. 1. A plot of 19 estimates of the rate of increase of the Earth's radius with time, based on the 19 data of Creer (1967) which give values for the radius of the Earth at various ages. The vertical axis gives the expansion rate in mm yr^{-1}, while the horizontal axis gives the time measured from the present back to the origin of the Earth (4.5×10^9 yr). The points are plotted with the ages of the data used to estimate the past radius. The open circles denote estimates based on palaeomagnetism, the closed circles denote estimates based on all other methods of calculating the past radius of the Earth. (Six points, numbers 2–4 and 16–18 inclusive of Creer (1967) represent estimates based on mean ages and mean ancient values of the radius respectively.) The palaeomagnetic data show a scatter diagram, and there are reasons for believing that this method of calculating the past radius of the Earth gives spurious results, especially where small amounts of expansion are concerned. Similarly, the non-palaeomagnetic point at (6.6 mm yr^{-1}, 0.25×10^9 yr) can be ruled out (see the text). The remaining 12 non-palaeomagnetic data define a reasonably coherent group indicating a mean rate of expansion over the history of the Earth of about 0.48 mm yr^{-1}.

indicating a large increase in the surface area of the globe since the Permian period (Carey, 1976; see also Steiner, 1977; Hilgenburg, 1962). A rate of expansion of about 6 mm yr^{-1} has also been suggested by Owen (1976) on the basis of related data to do with the drifting of the continents in the Mesozoic and Cenozoic, but such a high rate is unlikely for several reasons. Firstly, there is no reasonable mechanism for producing it (variable-G expansion can be expected to produce an increase in the Earth's radius according to $\dot{r}_E/r_E \propto -\dot{G}/G$, a relation used by Crossely and Stevens (1976) in a study of the possible expansion of Mercury). Secondly, it would imply such an enormous increase in the radius of the Earth over its history (assuming the rate to have applied at earlier epochs than the Mesozoic) as to conflict with established geophysical thinking about the temperature and structural evolution of the planet in the PreCambrian. Thirdly, although it will be seen that there are grounds for doubting the ability of the palaeomagnetic method to detect slow expansion (at rates of $\lesssim 0.6$ mm yr^{-1}), palaeomagnetism probably is accurate enough to rule out fast expansion (at rates $\simeq 6$ mm yr^{-1}). Fourthly, there is strong evidence against the simple fast expansion hypothesis from palaeogravity data (see below). These four points in conjunction effectively rule out the possibility that the Earth is undergoing a fast expansion.

Slow expansion is more viable, but despite the many estimates which have been made of the Earth's past radius which support expansion at a rate of 0.1–0.6 mm yr^{-1}, even the slow expansion hypothesis has been brought under attack on various counts (Wesson, 1978a). The most serious attack is that of McElhinny et al. (1978), who have used palaeomagnetism in the case of the Earth and a study of the morphology of the planetary surfaces in the other cases to obtain limits to possible expansion for the Earth, Moon, Mars and Mercury. They have concluded that the Earth has not increased its radius by more than 0.8% of its present value over the last 4×10^8 yr, and that the Moon has not expanded by more than 0.06% over the last 4×10^9 yr. Mars and Mercury may have changed their radii slightly over their histories, but if so, not by much. The Mercury data have also been used to set the limit $|\dot{G}/G| \lesssim 8 \times 10^{-12}$ yr^{-1} for variable-G cosmologies without continuous creation, and $|\dot{G}/G| \lesssim 2.5 \times 10^{-11}$ yr^{-1} in cosmologies like Dirac's ($\times$)-model where there is continuous creation. The palaeomagnetic limit for Earth expansion is equivalent to a limit on the expansion rate of about 0.13 mm yr^{-1}, while the limits on $\dot{G}/G$, if valid, rule out all three of the main variable-G theories discussed in Chapter 2. In view of the serious implications of the results of McElhinny et al. (1978) one must ask at this stage if palaeomagnetism as a means of estimating Earth expansion is really so reliable as it appears.

The palaeomagnetic method of estimating the radius of the Earth in the past has been criticized by Carey (1976), Wesson (1978a) and Smith (1978). Three criticisms of the method may be made, one general and two specific: (a) The palaeomagnetic methods discussed by Creer (1967) give rates of expansion which are scattered over a range from essentially zero up to about 6 mm yr^{-1}. This suggests that palaeomagnetism may not be capable of detecting moderate rates of expansion. (b) If the Earth is expanding, the deformation of the continents which one might expect to accompany the process would render the palaeomagnetic method unworkable (Carey, 1976). (c) Smith (1978) had drawn attention to the problem of deciding what central reference point one should use for the magnetic samples gathered from a given continental block. He has discussed this problem with especial regard to the palaeomagnetic work of Ward (1963; see also Ward, 1966), Van Hilten (1965; see also Van Hilten, 1968) and McElhinny et al. (1978). Ward (1963) has himself raised doubts about the palaeomagnetic method's ability to detect slow expansion, and has also (Ward, 1966) criticized the method of Van Hilten (1965) as being inconsistent. Smith's discussion concerns particularly the results of McElhinny *et al.* (1978), and he has concluded that low limits set on expansion by palaeomagnetism do not necessarily rule out the expanding Earth hypothesis in view of the ambiguities which exist in the application of the palaeomagnetic method.

Due to the questionable status of the palaeomagnetic expansion data and the unacceptability of the fast expansion hypothesis, a reasonable approach to the data of Figure 1 is to concentrate on the non-palaeomagnetic, slow expansion data (see the caption to Figure 1 and Wesson, 1978a). These data form a coherent group, and define a mean expansion rate of 0.48 mm yr^{-1} which has been approximately constant over the history of the Earth.

An expansion rate of approximately 0.5 mm yr^{-1} is noticeably higher than the rates predicted by variable-G cosmologies even with continuous creation (the highest rate for the scale-covariant theory with reasonable values of the cosmological parameters is about 0.3 mm yr^{-1}). However, a rate of about 0.5 mm yr^{-1} *is* of cosmological order (the radius of the Earth divided by 1×10^{10} yr defines a rate of 0.64 mm yr^{-1} and gives a rough measure of cosmological expansion on the geophysical scale). This indicates that the expansion has a cosmological origin even if the mechanism involved cannot be identified yet.

Expansion at the noted rate of about 0.5 mm yr^{-1} implies that the Earth was about half its present size at origin, 4.5×10^9 yr ago. This form of the expansion hypothesis has the desirable property of explaining the remarkable coincidence that the continents fit together neatly on a globe about half its present size (Creer, 1967; see also Hilgenburg, 1962; Runcorn, 1955). The implication is that the Earth has been expanding steadily at a rate of about 0.5 mm yr^{-1} since its origin, a process which led first to the cracking of an originally unbroken sial cover and subsequently to the pushing apart of the continents as the expansion continued. (This last process may have been assisted by convection inside the Earth, perhaps of the sort discussed by Kanesewich (1976) with a convection regime of the most fundamental type consisting of two main cells.) If expansion is real, it must have had notable effects on the geophysical evolution of the planet; but, despite the attractiveness of the expansion idea, one must be cautious about applying the hypothesis to the solving of geophysical problems.

An examination of the geophysical implications of the expanding Earth hypothesis has been given by Wesson (1978a; see also Wesson, 1973, 1975a).

To summarize the data discussed there, one can say quite simply that there is no evidence which rules out the slow expansion hypothesis but no evidence that proves the reality of expansion either. An old argument against expansion was that mountain building seems to imply contraction rather than expansion. This has led Lyttleton (1963a, b) to resurrect the idea of Ramsey (1949, 1950) that the core of the Earth is not composed of an iron-nickel alloy as believed by nearly all geophysicists but is instead a phase-changed silicate material, the mechanics of the phase-change process leading to a slow contraction of the Earth with time. This hypothesis, if true, would go against the expansion argument outlined above, so it is worth taking a closer look at it.

The phase-change hypothesis for the core was based originally on the fact that the compressibility k of silicate material was found to be a linear function of the pressure p at high pressures, and that geophysical data were consistent with a value of dk/dp that did not change appreciably at the boundary between the core and the mantle of the Earth (Ramsey, 1949; see also Ramsey, 1950). But the hypothesis in original form was unacceptable because it implied a solid silicate core (albeit with a metal-type structure of the phase-changed material), whereas seismic observations showed a fluid core. Lyttleton has re-examined the hypothesis with regard to mountain building and planetary structure for the Earth and Venus (Lyttleton 1963a, b respectively). In his form of the phase-change hypothesis, the Earth was initially all solid, but developed a phase-changed core that is liquid and is composed of silicate material with a metallic structure, and in which the atoms have had their outer shell electrons crushed off by the high pressure so that the material is also conducting (Lyttleton, 1965a). The slow development of the phase-changed region in the Earth is predicted to have led to a contraction of about 400 km in radius since the planet formed. The hypothesis has also been applied to the structure of Mars (Lyttleton, 1965b, c), Mercury and Venus (Lyttleton, 1969) and the terrestrial planets in general (Lyttleton, 1970, 1973, 1977). It is consistent with data on the secular accelerations of the Sun and the Moon, since the contraction leads to a small spin-up contribution to the Earth's rotation which tends to help in obtaining agreement between the total spin-down rates as inferred from theory and observation (Lyttleton, 1976). It is also consistent with the limits to changes of radius set by McElhinny *et al.* (1978), as shown by Lyttleton (1978a); but as seen above these limits are of doubtful meaning. Lyttleton's claims that the phase-change hypothesis is superior to the other theories of the cores and interiors of the terrestrial planets (Lyttleton, 1973, 1977, 1978a, b) are, moreover, meretricious. There are good reasons why the phase-change hypothesis has not been accepted by the majority of geophysicists, not the least important of which are the following two: (i) Mountain building does not require contraction but can instead be explained in a more natural way by plate tectonic processes perhaps working together with expansion (the account of the expanding Earth hypothesis by Dearnley (1966) pays particular attention to how expansion relates to mountain building); (ii) The cosmic abundance of iron is consistent with the hypothesis that the Earth's core is mainly metallic iron and not silicate (see Wesson, 1973, 1978a). There are other problems with the phase-change contraction hypothesis, and the arguments which have been made in favour of contraction cannot be regarded as significant criticisms of the expansion hypothesis.

However, the equations of planetary structure formulated by Lyttleton can be used in other contexts to examine the status of cosmological processes such as expansion and G-variability on the geophysical scale. Lyttleton and Fitch (1977) have criticized a calculation by Hoyle and Narlikar (1972) of the expansion of the Earth to be expected from a decreasing G, finding the amount of expansion to be only about 5 km in the last 2×10^8 yr for G-variability at the Dirac rate of $\dot{G}/G = -6 \times 10^{-11}$ yr^{-1}. Lyttleton and Fitch (1978a) have also criticized a calculation by Nordtvedt and Will (1972) of the increase in the moment of inertia of the Earth expected from a decreasing G; and they have criticized G-variability in general on the ground that the resulting increase in moment of inertia due to expansion of the Earth is in contradiction to the decrease which is implied by eclipse-derived data on the Earth's rotation, which indicate the presence of an accelerating term in addition to the (larger, mainly tidal) decelerating term (Lyttleton and Fitch, 1978a, b). These data, according to Lyttleton and Fitch (1978b) certainly imply $\dot{G} \simeq 0$ (and even imply $\dot{G} > 0$) for the iron core model of the Earth, although a $\dot{G} < 0$ term of size up to $|\dot{G}/G| \simeq 3 \times 10^{-11}$ yr^{-1} is possible for the phase-change model.

The criticisms of G-variability and expansion by Lyttleton and Fitch are not necessarily so negative as they appear. There are several unexplained problems connected with the deceleration of the Earth's rotation that need to be clarified before one can make definitive statements about the admissability of G-variability (Wesson, 1978a; Chapter 6); while the questions of how much expansion one can expect from a decrease in G, and whether the evidence in favour of expansion requires a cosmological mechanism to account for it, depend on assumptions about the internal structure of the planet. Beck (1969) has examined the energy requirements of expansion, finding that a change in radius by about 400 km over the history of the Earth can be accounted for solely by changes in gravitational potential energy within the planet. An extra 200–300 km can be accounted for by using radioactive energy, so the maximum expansion possible on the basis of conventional geophysical processes is about 700 km. (This estimate would have to be reduced to about 100 km if the moment of inertia of the Earth was higher in the past than it is now, as might be expected if the suggestion of Runcorn (1964) is valid and iron has been draining into the core from the mantle over geological time.) Changes in radius of about 1000 km can only be achieved on conventional grounds by adopting a highly unlikely density distribution for the Earth, and the amount of expansion expected on the basis of the most reasonable assumptions about the interior of the Earth is only about 100 km (Beck, 1961). These calculations of Beck make it obvious that large amounts of expansion require a cosmological process like G-variability to account for them. They confirm the inference drawn above that steady expansion at a rate of about 0.5 mm yr^{-1} with a connected change in radius of about 2000–3000 km since the origin of the Earth requires a cosmological driving force.

What exactly that driving force may be is a question that cannot yet be answered. As noted above, G-variability by itself cannot account for an expansion rate of about 0.5 mm yr^{-1}, and even G-variability in combination with continuous creation of matter can only account for expansion at rates of up to 0.3 mm yr^{-1} or so. One other mechanism which has been suggested by Van Flandern (1975) and Wesson (1975a, 1978a) is expansion without G-variability, due to space-time coupling of the cosmological expansion to systems of smaller scale. This can account for an increase in the Earth's radius at about 0.5 mm yr^{-1}, but it is a speculative hypothesis, a discussion of which is deferred until Chapter 6, where a cosmological basis for it is proposed. One way of testing both G-variability, matter creation and expansion (due to whatever cause) is by measuring the past value of the acceleration due to gravity at the Earth's surface ($g_E = GM_E/r_E^2$, where M_E is the mass of the Earth; see Wesson, 1978a and Chapter 6). Stewart has set a limit to changes in palaeogravity since the late PreCambrian by analyzing the pressure-temperature relation of upper mantle kimberlite nodules (Stewart, 1978), finding that g_E (now) $\lesssim g_E$ (PreCambrian) $\lesssim 2g_E$ (now). This limit confirms his earlier one of g_E (now) $\lesssim g_E$ (Phanerozoic) $\lesssim 1.4g_E$ (now) obtained from the non-occurrence of lawsonite in deep sedimentary basins (Stewart, 1977). As Stewart (1978) has noted, these limits rule out the fast expansion hypothesis for the Earth if G and M_E have not changed significantly over the last 1×10^9 yr; and while the slow expansion hypothesis survives, it must nevertheless be concluded that there is no evidence to suggest that g_E has changed significantly during the last 1×10^9 yr or so.

This may mean that conventional theory is correct and that G, and the mass

M_E and radius r_E of the Earth have not changed in the geological past. But it may also mean that there is a conspiracy effect in which G, M_E and r_E vary in a manner that results in g_E remaining approximately constant. Clearly, what is needed is more geophysical data on the possible variation of the parameters involved. Combined with limits from other sources on variations in G and the masses of bodies, such geophysical data would enable the question of the expansion of the Earth and its cause to be answered with some degree of certainty.

CHAPTER 6

DISCUSSION

6.1. Introduction

Having examined variable-G theories, gauge-invariant particle theories and various aspects of astrophysics and geophysics which involve gravitation and particle physics, it is the purpose of the present chapter to discuss a little more fully how these topics relate to each other and to examine some subjects which affect all of the aforementioned fields. The two main subjects in the latter class are gravity experiments and the type of scale invariance known as self-similarity.

Gravity experiments as they concern particular variable-G theories or particular aspects of astrophysics and geophysics have been mentioned in the relevant chapters above. One can recall that all three of the main variable-G theories predict values of the rate of change of G in the range $10^{-11} - 10^{-10}$ yr^{-1}. The theories of Hoyle/Narlikar and Canuto *et al.* (in general form) can be adjusted to give as low a rate of change of G as desired, but the Dirac theory as it employs the Large Numbers Hypothesis predicts a value of $\dot{G}/G$ which is fixed to a certain degree by the age of the Universe ($\dot{G}/G \simeq -6 \times 10^{-11}$ yr^{-1}). The same concerns the theory of Canuto *et al.* in the form in which it uses the Large Numbers Hypothesis as a gauge-choosing condition. There are already indications from astrophysics that $|\dot{G}/G| \lesssim 1.5 \times 10^{-12}$ yr^{-1} and from geophysics that $|\dot{G}/G| \lesssim 2 \times 10^{-11}$ yr^{-1}. To the extent that these limits depend on ancillary assumptions about astrophysical and geophysical processes, one can argue that they do not provide *direct* evidence against G-variability at the rates expected from theory. The crucial question then becomes: What evidence is there from (more or less) direct experiments for or against a time-variation of G? The object of Section 6.2 is to review the more direct tests of G-variability, both those which have been carried out and those which are as yet only planned, and to discuss how such experiments relate to the result of Van Flandern (1975) which indicates a finite rate of change of the gravitational parameter.

Scale invariance of the type known as self-similarity is important because it provides a technical way of introducing such invariance into already-established theory (both gravitation theory and particle theory, though it has been most used in the former context). Self-similarity can also be compatible with new theories of physics, and it has special relevance to gauge theories of gravity that can be formulated without the presence of unnecessary dimensional parameters. While the technique of self-similarity has been used in various branches of physics for many years, it is only recently that its implications for cosmology and gravitation have begun to be realized. The technique is therefore still largely new in the field of gravitational theory, and for this reason considerable attention is given in Section 6.3 to examining both the meaning and application of the self-similar type of scale invariance.

6.2. Gravity Experiments

The reasonably direct experiments which have been carried out or are planned to measure the variation of G with time can conveniently be classified into those that employ radar methods and those that employ laboratory methods.

The radar methods are based on the fact that if G changes with time in a system composed of a parent body and an orbiting secondary body, the distance between the two bodies also changes with time. By monitoring distances in the Solar System, and comparing the motions of its component bodies with those predicted by general relativity, it is possible to put limits on systematic motions of expansion or contraction, and so obtain limits on the variation of G. This method is reasonably direct. The only possible problem with it as a way of obtaining a limit on $\dot{G}$ is that secular mass loss from the bodies concerned can have the same effect as a secular decrease in G (Wesson, 1978a). Fortunately, the bodies of the Solar System are not being seriously affected by secular mass loss on time scales of the order of 10^{10} yr, as far as can be judged. This means that any departures from general relativity of the type being discussed can be ascribed to a time-dependent G; or possibly, as pointed out by Van Flandern (1975) and Wesson (1978a), to an expansion that is unconnected with a change in G.

The two-body problem with variable G has been studied theoretically by Vinti (1974) and Hut and Verhulst (1976; see also Bishop and Landsberg, 1976; Maiti, 1978). The corresponding problem with variable masses has been treated by Verhulst (1975). The application of radar methods to the various two-body systems formed by the Sun and the planets was first made by Shapiro et al. (1971), who obtained the limit $|\dot{G}/G| \lesssim 4 \times 10^{-10}$ yr^{-1} by tracking of planetary orbits. This limit has been improved on, and radar data show that $|\dot{G}/G|$ is effectively zero to within limits of $4 - 33 \times 10^{-11}$ yr^{-1} depending on which planet is employed in the determination, the noted figures being standard deviations of the most and least accurate observations

(Reasenberg and Shapiro, 1978; see also Reasenberg and Shapiro, 1976). Irrespective of how one analyzes the data, these measurements show that G is certainly not varying at more than $|\dot{G}/G| \approx 10^{-10}$ yr^{-1}. This conclusion has been confirmed by Anderson *et al.* (1978), who have reviewed tests of general relativity using astrometric and radio-metric observations of the planets, and state that all the available data indicate that $|\dot{G}/G| \lesssim 1.4 \times 10^{-10}$ yr^{-1}.

Other possible ways of testing general relativity using methods involving the planets or experiments in space have been discussed in the volume edited by Wrigley (1978), and of the data discussed there the most pertinent result as regards $\dot{G}/G$ is that of Anderson *et al.* (1978) just quoted. Radar experiments involving the planets are important because these methods (which use the radar principle of reflection of electromagnetic waves of various wavelengths to estimate distances) are best suited to measuring the deviations from Newtonian theory that are characteristic of general relativity and related theories. The deviations involved are small by astronomical standards, the general relativistic contribution to the orbit of Mercury being about 2 km, for example (Clemence, 1962). A related set of experiments which employ the radar principle involve the bouncing of a laser beam off a reflector on the Moon. Various applications of lunar laser ranging have been treated in the book edited by Mulholland (1977) and its use in testing cosmological and geophysical hypotheses has been discussed by Wesson (1978a). Among other uses, this technique can be employed in connection with the determination of Universal Time (Stolz *et al.*, 1976; King, 1978) and in testing some $\dot{G}$ theories (Williams *et al.*, 1976; Shapiro *et al.*, 1976). It is, in fact, the laser tracking experiments which have effectively ruled out scalar-tensor theories of gravity like the Brans/Dicke theory as viable accounts of gravitation (see Chapter 8). Radar methods, whether using planetary or lunar reflections, can thus be used to examine the status of non-Einsteinian theories of gravity once the parameterized post-Newtonian approximations to such theories have been worked out so that it is possible to compare their predictions with observation via the use of Newtonian ephemerides. This aspect of such methods has been discussed by Nordtvedt (1977), who has noted that gravity experiments in general show no departures from Einstein's theory to an accuracy of better than 1 percent.

The radar and laser methods have the attractive property that the longer they are used, and the more data are collected, the better is the accuracy of the resulting measurements. They will therefore continue to be developed, possibly in combination with artificial-satellite technology. The problem of two fixed centres and a time-variable G has been considered in relation to artificial Earth satellites by Bekov and Omarov (1978); while Hughes

(1977a, b) has suggested that the use of microwave ranging techniques like those employed in the Viking lander project on Mars could in principle enable the Earth-Mars distance to be monitored to an accuracy of 15 cm with 4 years' data, so enabling a direct test to be made of Dirac's Large Numbers Hypothesis. Writing in a book edited by Halpern (1978a) on methods of measuring cosmological variations of G, Dirac (1978) has reviewed the Large Numbers Hypothesis (see Chapter 2), which with an age of the Universe inferred from astrophysical data implies $\dot{G}/G \simeq -6 \times 10^{-11}$ yr^{-1}. A value of this size for $\dot{G}/G$ is already dangerously close to being impeached by the radar experiments discussed above; and as will be seen below, such a value for $\dot{G}/G$ will be either disproved or confirmed by laboratory experiments which are planned to measure possible changes in G and/or changes in the masses of bodies at rates of $10^{-11} - 10^{-12}$ yr^{-1}.

Laboratory methods that have been proposed to measure $\dot{G}/G$ are of various sorts. Braginskii and Ginzburg (1974) have suggested two ways of measuring $\dot{G}/G$ to an accuracy of about 1×10^{-11} yr^{-1}. One involves a pendulum experiment and the other involves the use of a spring-loaded gravimeter of the type used in geophysics, the object being to obtain $\dot{G}/G$ via the intermediate step of estimating secular changes in the value of the local acceleration due to gravity g_E at the Earth's surface. (It is interesting to note that $\dot{G}/G$ can actually be determined to better accuracy than the absolute value of G, and the lack of an accurate determination of this parameter has led Stacey (1978) to suggest a way of determining G with improved exactitude by measuring g_E using an apparatus floating at depth in the sea.) Instrumentally, $\dot{g}_E/g_E$ can in principle be measured to an accuracy of about 1×10^{-11} yr^{-1} and $\dot{G}/G$ can be obtained to comparable accuracy given a connection between g_E and G. The main problem with this method is that g_E changes on short time scales due to purely geological causes (Wesson, 1978a). Braginskii and Ginzburg have noted that processes to do with continental uplift and subsidence imply changes in g_E at rates of $\dot{g}_E/g_E \simeq 10^{-8}$ yr^{-1}, which must somehow be averaged out. Despite this problematical aspect, the test is of added interest because it also provides in theory a way of testing for a possible secular expansion of the Earth (Chapter 5), since if one uses the usual formula $g_E = GM_E/r_E^2$ (where M_E and r_E are the mass and radius of the Earth), then $\dot{g}_E/g_E = -2\dot{r}_E/r_E$ if G is constant. Thus, if other experiments to be described below show that G is not varying significantly with time, the possibility exists of measuring changes in the radius of the Earth at levels of $|\dot{r}_E| \simeq (10^{-11}r_E/2)$ yr$^{-1} \simeq 0.03$ mm yr^{-1}. This is ample accuracy for testing the expanding Earth hypothesis.

Laboratory experiments capable of measuring $\dot{G}/G$ to an accuracy of about 1×10^{-11} yr^{-1} have been discussed further by Braginskii et al. (1977; see also Davies, 1977c). They have noted that such experiments can also be used to test other consequences of relativistic theories of gravity, such as the contribution of stress to mass in theories like general relativity and possible effects of the uneven mass distribution of the Galaxy on gravity in the laboratory. Similarly, Halpern and Long (1978) have proposed two methods for measuring a possible time-variation of G based on the use of superconducting technology. However, gravity experiments like these and the ones hitherto described have yet to be carried out, and while they may prove to be of use, most attention has been focussed recently on a pair of experiments which have already been developed to a point at which it is clear that measurements of $\dot{G}/G$ are quite feasible even down to a level of order 10^{-12} yr^{-1}.

The University of Virginia group has designed and developed two experiments which are related in their technological aspects but are aimed at testing different non-Einsteinian processes. One is an experiment to measure $\dot{G}/G$ and the other is an experiment to measure a possible change of mass on a secular time scale. The latter phenomenon is predicted by several theories of gravity (Wesson, 1978a), and the effort to detect it constitutes in some ways the more interesting experiment since it holds out the hope of showing if the much-discussed process of continuous creation of matter is viable or not. The matter-creation experiment aims to measure the quantity $\dot{m}/m$ for the rate of change of a test mass m. The University of Virginia experiments make use of recent advances in the design of drive mechanisms for keeping a turntable rotating at constant speed (Gillies et al., 1978), and Figure 2 shows how an apparatus consisting essentially of two rotating cylinders can be employed to detect a change of mass of one cylinder with respect to the other. The experiment has been described in detail by Ritter et al. (1978), and it is expected that eventually a decay time for the system of order 10^{18} sec or better will be achieved. This is long enough to enable continuous creation of the type predicted by the Dirac theory (Chapter 2) on a time scale of order 10^{10} yr $\simeq 3 \times 10^{17}$ sec to be detected.

The University of Virginia $\dot{G}/G$ experiment is similar in design to the $\dot{m}/m$ experiment. It has been described in detail by Ritter and Beams (1978), and consists essentially of an isolating table, a drive mechanism and a computer-controlled angle sensor (as in Figure 2), interacting with a rotating turntable on which are mounted a pair of large masses and a pair of small masses. The force of attraction between the small masses and the large masses is counterbalanced by a drag mechanism. A decrease in G would manifest itself as a change in the force of attraction between the large and small masses, and

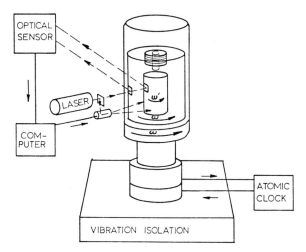

Fig. 2. A diagram of the continuous creation experiment of Ritter et al. (1978) being carried out at the University of Virginia. Two cylinders of temperature-stable ceramic (Zerodur) rotate concentrically in an evacuated region inside an acoustic and magnetic shield. The inner cylinder is magnetically suspended from the outer one which rotates with precise angular velocity ω. Mass created in the inner cylinder tends to slow it down. A feedback system employing laser pulse sensing and photon driving keeps the inner cylinder velocity ω' very near to ω. The forward/backward asymmetry needed in these feedback-driving pulses to keep $\omega' = \omega$ constitutes the signal. With the two cylinders running synchronously, viscous, magnetic hysteresis and other damping effects are kept near zero. (Reproduced with the permission of G. T. Gillies of the University of Virginia.)

so could be detected via the drag mechanism. This experiment has been designed to measure $\dot{G}/G$ to an accuracy of order 10^{-11} yr^{-1}, but with improvements in technology it may be possible to reach an exactitude of a few parts in 10^{12} yr^{-1}.

The direct experiments described above, both the ones which have been carried out and the ones which are planned, make it clear that theoretical grounds for believing in G-variability at rates in the range $|\dot{G}/G| = 1 \times 10^{-11}$–$1 \times 10^{-10}$ yr^{-1} are in a precarious state. Experimental evidence is clearcut in showing that G is not varying at rates $|\dot{G}/G|$ larger than about 1×10^{-10} yr^{-1}, and there is evidence from radar experiments and from geophysics that $|\dot{G}/G|$ is probably not larger than 4×10^{-11} yr^{-1} or 2×10^{-11} yr^{-1} respectively. The question naturally arises of how one is to reconcile these limits with the one outstanding result which does indicate a finite value for $\dot{G}$, namely that of Van Flandern (1975) which is $\dot{G}/G \simeq -8 \times 10^{-11}$ yr^{-1}.

Van Flandern's way of determining $\dot{G}/G$ is based on finding that part of the secular acceleration of the Moon which is not due to geophysical effects such as tides. This is done by monitoring the Moon's position in its orbit using the very accurate method of observing lunar occultations of stars. The occultations are timed directly against the Atomic Time service, which gives a direct time standard that is equivalent to the atomic time of the theories of Dirac and others (as opposed to Ephemeris Time, which is an essentially gravitational time scale used previously in astronomy and based on the motions of the planets). A preliminary estimate of the secular acceleration of the Moon on Atomic Time was given by Van Flandern (1970). A fuller account yielded the noted result $\dot{G}/G \simeq -8 \times 10^{-11}$ yr^{-1}, which Van Flandern realized could also be interpreted in terms of the equivalent processes of mass loss or expansion with G constant (Van Flandern, 1975; see also Wesson, 1978a). As mentioned previously, mass loss in the Earth-Moon system can be shown to be negligible, and in the absence of a theory predicting a simple expansion of space-time, Van Flandern interpreted his result in terms of G-variability (Van Flandern, 1975, 1976). In an update of the status of the lunar-occultation $\dot{G}$ determination, Van Flandern (1978) adjusted the value of $\dot{G}/G$ slightly to $\dot{G}/G \simeq -6 \times 10^{-11}$ yr^{-1} approximately, a value which is also in agreement with data on transits of Mercury and is compatible with the multiplicative ($\times$)-form of the Dirac theory. (However, in the interpretation of Van Flandern's result given in Section 6.3, the straightforward lunar occultation data with the associated value of $\dot{G}/G \simeq -8 \times 10^{-11}$ yr^{-1} will be adopted.) Further refinement of the result is to be hoped for as more data and data of better quality become available.

A value of $|\dot{G}/G| = 6\text{--}8 \times 10^{-11}$ yr^{-1} is in apparent contradiction with other, indirect methods of estimating $\dot{G}/G$. One would like if possible to know the reason for the discrepancy, and to see how Van Flandern's result compares with other, less direct methods of estimating $\dot{G}/G$ by using the Earth-Moon system. Some of the most recent results on $\dot{G}/G$ from studies of the rotation of the Earth are those of Blake (1977, 1978). These are based on an application of a method of Van Diggelen (1976), which was applied in original form to the problem of obtaining a limit on possible Earth expansion during the last 5×10^8 yr. Van Diggelen noted that if P_E, r_E and ω_E are the angular momentum, radius, and angular velocity of rotation of the Earth respectively, then $\dot{P}_E/P_E = 2\dot{r}_E/r_E + \dot{\omega}_E/\omega_E$, assuming that G is constant. Data are available on $\dot{\omega}_E$ and $\dot{P}_E$, so it is feasible to evaluate $\dot{r}_E/r_E$ and so obtain a limit on Earth expansion. Values of $\dot{\omega}_E$ and $\dot{P}_E$ can be obtained from the growth lines of fossil corals, the value of $\dot{\omega}_E$ coming from the number of days in the year in past geological periods and the value of $\dot{P}_E$ coming from the number of days in the synodic month in the Devonian period. There are some assumptions involved in the reducing of the raw fossil data to a form suitable for use in the quoted equation, perhaps the most important of which is the assumption that the angular momentum of the Earth-Moon system has been constant in the past. Further, Van Diggelen used only one data point in estimating the ancient value of P_E (i.e., he used only the Devonian and present values of the angular momentum to obtain $\dot{P}_E$), and ignored any possible change in the moment of inertia of the Earth due to effects other than a simple expansion. For these reasons his conclusion that $\dot{r}_E/r_E$ has been effectively zero over the past 5×10^8 yr is not to be taken too seriously (compare the discussion of Earth expansion in Chapter 5). The principle of the method is valid, though, and in its application by Blake (1977, 1978) it has yielded close limits on a possible change in G.

Blake's method is essentially to use a more complicated, and more general ($\dot{G} \neq 0$) version of the approach employed by Van Diggelen. With values of the various parameters involved calculated from the lengths of the year and month as recorded by fossil bivalves and corals of ages up to about 5×10^8 yr, Blake (1977) has found $\dot{G}/G = -0.5(\pm 2) \times 10^{-11}$ yr^{-1}. By employing the method in conjunction with the predictions of the Large Numbers Hypothesis of Dirac, Blake (1978) has also concluded that the simple $G \propto t^{-1}$ theory (with no matter creation) and the $(+)$-form of Dirac's cosmology can be rejected, but that the $(\times)$-form is acceptable, as is the conventional theory in which $\dot{G} \equiv 0$ and there is no creation of matter. The analysis of Blake (1978) takes properly into account a possible expansion of the Earth in response to a decreasing G, and shows quite clearly that a time-variable G is only compatible with data on the rotation of the Earth if G-variability is coupled with continuous creation of the type in which matter is created where matter is already most dense.

While a simple $G \propto t^{-1}$ dependency in the Earth-Moon system is not compatible with Van Flandern's finite result for $\dot{G}/G$, one sees from preceding comments that a decrease in G at a rate of $\dot{G}/G \simeq -8 \times 10^{-11}$ yr^{-1} might be compatible with the dynamics of the system if it occurs in combination with other processes (such as continuous creation or structural reorganization inside the Earth or a departure from the law of conservation of angular momentum over long periods, all of which would tend to offset the effects of a decreasing G). It is difficult to test such processes using the direct methods discussed above. Lunar laser-ranging data by themselves are not extensive enough to discriminate between the zero G-variation rate of conventional theory and G-variation at the rate found by Van Flandern (see Calame and Mulholland, 1978). This means that one is obliged to rely on the less direct methods involving the rotation of the Earth which have already been mentioned. In this regard, it must be stated that there are still apparently unexplained tendencies in the long-term behaviour of the rotation of the Earth that can mask $\dot{G}$ effects. Part of the discrepancy which used to exist between different methods of estimating the secular spin-down rate of the Earth can in all probability be attributed to the falsification of observations of ancient eclipses by the people who claimed to have experienced them, and to differences in the ways in which the records have been analyzed (Newton, 1977; see also Stephenson and Clark, 1978). But there is also evidence for real changes in the regular despinning of the planet's rotation which probably have their origin in the operation of non-gravitational forces in the Earth-Moon system and which manifest themselves as surges of acceleration that cannot be accounted for in terms of known geophysical processes (Newton, 1972; Wesson 1978a; Kahn and Pompea, 1978; see also Mansinha *et al.*, 1970). Such effects make the problem of estimating $\dot{G}/G$ from the rotation of the Earth a difficult task.

In spite of the many things which might lead to apparent contradictions of analysis, Muller (1978; see also Muller, 1975) has found that several astronomical data do have a consistent interpretation which is compatible with the result of Van Flandern (1975). Muller has noted that a range of astronomical data having to do with eclipses, lunar occultations of stars, transits of Mercury and solar equinoxes are all in agreement with the belief that the non-tidal contribution to the spin-down of the Earth is approximately zero (i.e., that the part of $\dot{\omega}_E/\omega_E$ which is not due to tides and related processes is insignificant). By combining this assumption with the value for the total $\dot{\omega}_E/\omega_E$ ($\simeq -2.5 \times 10^{-10}$ yr^{-1}) as calculated from data on timed ancient solar equinoxes, Muller has found that the rotation of the Earth is compatible with a decreasing value of G at a rate of $\dot{G}/G = -6.9(\pm 3.0) \times 10^{-11}$ yr^{-1}. This result is independent of the motion of the Moon and of Van Flandern's result. It represents formally an independent estimate of G-variability, but its essential significance is that it demonstrates that other data connected with the rotation of the Earth are consistent with Van Flandern's occultation-derived value of $\dot{G}/G \simeq -8 \times 10^{-11}$ yr^{-1}.

One concludes that as far as the rotation of the Earth and the dynamics of the Earth-Moon system are concerned, Van Flandern's outstanding finite determination of a changing G may be valid. However, in view of the investigations of G-variability at a level of $\dot{G}/G \approx 10^{-12}$–$10^{-11}$ yr^{-1} which one can expect from laboratory experiments, it is worthwhile to consider the situation in which Van Flandern's result remains valid but in which no other evidence for G-variability can be found. This situation is not at all unlikely to be realized. Indeed, some workers would say that this situation already exists in view of the stringent limits on $\dot{G}/G$ which are set by astrophysical and other methods of testing for a time-dependence of gravity. Accordingly, in the next section, a hypothesis will be considered which might explain Van Flandern's data without invoking G-variability, such a hypothesis being one aspect of a possible scale-invariant theory of gravity.

6.3. Scale Invariance and Self-Similarity

Scale invariance, in one form or another, has been used in connection with theories of gravitation of the Weyl type (Chapter 2) and the scalar-tensor type (Chapter 8), and in various other thories of gravity (Chapter 9). It has also been used in particle physics (Chapter 3). Nearly all of these accounts, as they have to do with gravity, depart from Einstein's theory, although there are modifications of the latter which are scale invariant (Chapter 8). However, it is clearly of interest, in view of the established status of Einstein's theory as opposed to the doubtful status of other theories, to ask: Is there a way in which general relativity as it stands can be made scale invariant? The answer to this question is in the affirmative insofar as there is a restricted form of scale invariance which is compatible with Einstein's theory in unmodified form.

Self-similarity (or similarity) is a term which means physically that arbitrary scales of length or time are absent from, say, moving matter distributions. The term means mathematically that the equations of motion of the system can be written in dimensionless form by a suitable choice of one dimensionless variable. Dimensionless quantities are invariant under scale changes of some types, and self-similarity itself is a kind of symmetry property which is observed to characterize hydrodynamical systems in which there are no arbitrarily-imposed boundaries. That is, it is an observed symmetry of Nature that fluids flowing in space remote from boundaries tend to have self-similar distributions, meaning that the density, pressure, velocity and gravitational potential have forms which are solutions of the dimensionless equations of motion and depend in essence on only one dimensionless variable.

There is a considerable literature on self-similarity. The books by Birkhoff (1950) and Sedov (1959) lay a foundation for self-similarity in Newtonian hydrodynamics. Other literature on astronomical problems which have been treated using Newtonian self-similarity can be found in Wesson (1978a, d). Recently, there has been renewed interest in Newtonian self-similarity in the astronomical field as it relates to galaxy formation (Press and Schecter, 1974; Doroshkevich and Zel'dovich, 1975; Silk and White, 1978), blast waves (Newman, 1977; Lerche, 1978a, b) and critical points (Gitterman, 1978). Self-similarity in general relativity has been studied by Cahill and Taub (1971), Henriksen and Wesson (1978a, b) and Wesson (1978a, 1979a, b).

Problems which have been investigated using relativistic self-similarity concern black hole accretion (Carr and Hawking, 1974; Lin et al., 1976; Bicknell and Henriksen, 1978a, b) rotating disks (Wesson and Lermann, 1978a; Lynden-Bell and Pineault, 1978a, b), cosmological models (Wesson, 1978d, e, 1979a) and symmetry properties such as the Cosmological Principle which might be formulated in terms of self-similarity (Wesson, 1978c, 1979c). Although the physical concept involved in these various treatments is the same (viz: the scale-free description of fluid systems), the mathematical approaches that have been employed differ somewhat. As far as general relativity is concerned, the simplest mathematical formalism for self-similarity is that of Henriksen and Wesson (1978a), in which the metric coefficients g_{ij} and the essential parts of the functions describing the matter properties are dependent only on the combined dimensionless variable ct/R (where t is a time coordinate and R is a distance coordinate). This variable is a dimensionless ratio of two lengths and is invariant under scale transformations that preserve the dimensionality of the numerator and denominator. (Actually, transformations that are not of this type, such as coordinate transformations that alter the functional form of t and R, cannot be admitted in a self-similar description because there are no intrinsic time or length scales which can be employed to render such transformations dimensionally consistent.) It is possible to find cosmological models which are self-similar in the sense that the metric coefficients are functions only of ct/R (Henriksen and Wesson, 1978a). It would be of interest if such a model could be found that is in agreement with the observed Universe, since in that case one would have a scale-invariant cosmology without going outside the structure of Einstein's theory of general relativity.

A model of this type was given by Wesson (1978d). Its observable properties were evaluated in Wesson (1979a), the main one of which concerns the number density of high-redshift astronomical sources such as QSOs. This property was examined in detail in Wesson (1978e). The model as a whole has several peculiar features. It is globally inhomogeneous, the size of the inhomogeneity at any epoch depending on the ratio of ct/R to a certain constant α, the inhomogeneity being small when $ct/\alpha R$ is large. This means that the inhomogeneity becomes less and less as the age of the model increases, and in the limit $t \to \infty$ the matter density ρ becomes homogeneous. This property of beginning inhomogeneous and tending to homogeneity was also present in a similar inhomogeneous model of Bonner (1972). He showed in fact that all pressure-free, parabolic, spherically-symmetric solutions to Einstein's equations tend to become identical to the isotropic Einstein/de Sitter solution in the $t \to \infty$ limit (Bonner, 1974). The self-similar model is pressure-free and parabolic, so its behaviour fits into the scheme studied by Bonner. For epochs t which are finite, the density ρ is inhomogeneous to a greater or lesser degree depending on the value of the variable ct/R. There is a related anisotropy in Hubble's parameter H, and anisotropies are also present in the 3 K microwave background at some level due to the inhomogeneous distribution of galaxies in the model. These two anisotropies can be identified with a known anisotropy in the redshift distribution of the galaxies and with an apparent peculiar velocity of the Galaxy with respect to the 3 K background (see Chapter 11). The model avoids the well-known causality problem presented by the fact that the temperature of the microwave background in the Friedmann models is nearly the same in regions of space-time that were not causally connected (i.e., were not within each other's horizon) in the early Universe. This contretemps is avoided in the self-similar model by the peculiar circumstance that in its early stages the space part of the metric is not of the usual three-dimensional type. (Mathematically, near $t = -\alpha R/c$ where α is a constant, the azimuthal interval tends to zero while the radial interval tends to infinity.) It was pointed out by Saslaw (1977) that the isotropy of certain models of the Universe can in principle be attributed to the fact that they pass through pathological phases when the space part of the metric changes its dimensionality, being for example two-dimensional at some early epoch before later evolving to the more familiar three-dimensional form. In this way one can understand the present near isotropy of the 3 K background. A comparison of the small anisotropy in the Hubble parameter, which the self-similar model predicts, with the observed velocity field of the galaxies shows that the model is in dynamical agreement with observation for values $ct_0/\alpha R \gtrsim 10$ of the dimensionless variable at the present epoch. Indeed, it can be shown (Wesson, 1979a) that a value of $ct_0/\alpha R \gtrsim 10$ ensures that the model is in agreement with all the available data on Hubble's parameter, the deceleration parameter, the magnitude/redshift (or Hubble) diagram, the number density/redshift relation and the microwave background.

The question of whether a model of the self-similar, scale-invariant type is *better* than one of the conventional, non-invariant models is something that must be decided by a comparison with specific observations. As mentioned above, the observational status of the self-similar model depends most crucially on the predicted versus observed number densities of astronomical sources at high redshifts. This observational aspect of the model (Wesson, 1978e, 1979a) is the most striking way in which it differs from the isotropic Friedmann solutions to Einstein's equations, which latter are usually taken as the standard models in cosmological studies. The self-similar model predicts that the number density of all types of source should appear to increase away from a typical observer, roughly as $(1 + z)^p$ with $1.5 \lesssim p \lesssim 5.2$, where z is the redshift. There is some evidence that normal galaxies, Seyfert galaxies, N-galaxies, radio galaxies and QSOs form a series, in which a part at least of the observed 'pile-up' of sources at high redshifts is due to evolutionary effects intrinsic to the sources (see, for example, Rowan-Robinson (1977a) and Rees (1977a); and Stannard (1973), Setti and Woltjer (1973) and Netzer et al. (1978) for discussions of the Hubble diagram for QSOs). However, while part of the pile-up could be due to evolution intrinsic to the sources, it seems likely that a major part could also be due to the cosmology, because the observed number density of sources rises very steeply at high redshifts as $(1 + z)^p$ with $p = 5.5$–6.0 approximately (Longair, 1966; Schmidt, 1968, 1970, 1974, 1975; Rees, 1971; Green and Schmidt, 1978; see also Just, 1959; Rowan-Robinson, 1972a, b; Rees, 1972b; Von Hoerner, 1973a; Osmer and Smith, 1977; Stewart and Hawkins, 1978). Thus, the self-similar model is in agreement with the number density of high-redshift sources, although the data concerned are to a certain degree ambiguous in interpretation.

In addition to the mean rise in number density of sources with z, the model also predicts that there should be an anisotropy in the number density of high-z sources that depends on direction, the size of the anisotropy depending on ct_0/R and being small if that variable is large. The space distribution of QSOs is usually taken as being uniform over the sky. Plagemann (1973) found that there was no gross anisotropy in the angular distribution of QSOs, while Wills and Ricklefs (1976) found that there was no significant anisotropy in QSO redshift distribution with respect to the two Galactic hemispheres. Setti and Woltjer (1977) found no evidence of clustering of QSOs, in particular pairing. This conclusion was confirmed by Wills (1978), who reviewed the evidence for and against clustering, pairing and large-scale anisotropies in the distribution of quasars. However, as Wills pointed out, it is impossible to make a case against there being global anisotropies in the QSO distribution of the type expected from inhomogeneous cosmological models, because uniform surveys of quasars have not yet been made over areas of sky large enough to permit the effects involved to be detected with certainty at the expected levels. Despite this, there is some evidence in favour of anisotropy. Firstly, Prakash (1978) has deduced from a statistical analysis of quasar z-values that QSOs are clumped on a scale of order 10^3 Mpc (the horizon of the Universe is of size $ct_0 \simeq 6 \times 10^3$ Mpc if $t_0 \simeq 2 \times 10^{10}$ yr). Secondly, a study of a new, near complete sample of quasi-stellar radio sources from the 4 C catalogue by Wills and Lynds (1978) has revealed a deficiency in the number of QSOs with Galactic coordinates $|b| \geq 20°$ in the section of the sky between 12 hr and 24 hr R.A. compared to the number of QSOs in the section of sky between 0 hr and 12 hr R.A. In view of the comment made above, neither of these two studies can be taken as proving that high-redshift sources are distributed in the manner expected from an inhomogeneous cosmological model, but they show nevertheless that effects of the type expected from such a model may be present in the data.

To summarize this discussion of the self-similar model, it is apparent that scale-invariance of a natural type (viz: the absence of arbitrary scales) can be successfully included in Einstein's theory in unmodified form. There is at least one such model that agrees with the observations, at any rate to the extent to which the latter depend on interpretation. The existence of a model of this type should lead us to ask if it is justifiable to construct theories of gravity that go outside general relativity on the ground that the theory as a whole (as opposed to just one model) should be scale-invariant. The answer to this depends to a certain extent on what type of scale invariance one is interested in (various meanings of this term are noted and discussed in Chapters 8 and 9). But at a more fundamental level the question is one of philosophy, rather than something that depends on the technicality of a definition. At this level one can make the following tentative comment: If scale-invariance (of a given type) is important, then it is desirable that the whole theory be constructed so as to be scale invariant, rather than seeking specific scale-invariant solutions of a theory that does not itself recognize the importance of that property as a postulate.

To give an example connected with the subject matter of preceding paragraphs: Einstein's equations by themselves are not in general compatible with the property of self-similarity, even though it is possible to find special solutions to these equations that are self-similar (Wesson, 1979b). Clearly, if one regards self-similarity as being itself of fundamental importance, all solutions of a given set of field equations ought to be self-similar. Theories of wider scope than Einstein's are therefore justified. In particular, gauge theories, in which the field equations can take on the same form as Einstein's equations given a certain choice of the gauge function, are good candidates for theories of gravity. They ensure that the theory shall be in agreeement with the classical tests of general relativity (because there is a gauge in which the equations are the same), while also providing an opportunity to widen the scope of the theory in other directions, and especially in directions which involve astrophysics and cosmology.

Some clues to a possible gauge theory of gravitation of wider scope than Einstein's theory were noted by Wesson (1978a; see also Wesson, 1973, 1975a, 1977a, 1978b; and Chapter 10). There are three separate and apparently unrelated things which affect geophysics and astrophysics but which can all be understood on the basis of the hypothesis that there exists a gauge theory of gravitation in which the cosmological expansion couples to bodies of small size. This hypothesis can be formulated in a definite way by using an extension of the conditions for metric continuity which hold in Einstein's theory. (The general relativistic conditions and instances in which they have been used in astrophysics were reviewed in Wesson (1978a) and Wesson and Lermann (1978a); see Kantowski (1969) for a typical application to a cluster of galaxies.) That is, one can formulate the hypothesis of space-time coupling as a condition on the continuity of the metric coefficients g_{ij}, where the precise form of the g_{ij} will depend on the gauge theory concerned and on the cosmological model being used (Wesson, 1978a). In this way one can mathematically describe a system in which all bodies expand at the Hubble rate $\dot{r} = Hr$, where H is Hubble's parameter with its cosmological value ($H \simeq 75$ km sec^{-1} Mpc$^{-1} \simeq 2.5 \times 10^{-18}$ sec^{-1}). This may sound rather drastic, but the effects of this hypothesis on scales that are not of cosmological size are only of the same order as those of the processes characteristic of the theories treated in Chapter 2. One is dealing with relatively minor effects that proceed very slowly at a rate of order $H \approx 10^{-11}$–10^{-10} yr^{-1}. Despite this, there are some salient consequences of the coupling hypothesis, and in particular it explains the following three facts:

(a) The result of Van Flandern (1975) that $\dot{G}/G \simeq -8 \times 10^{-11}$ yr^{-1} as indicated by occultations of stars by the Moon is dynamically equivalent to an expansion of the lunar orbit at a rate given by $\dot{r}_M/r_M = -\dot{G}/G = 2.6 \times 10^{-18}$ sec^{-1}. The remarkable coincidences between this figure and the size of H can be understood on the basis of an expansion of the lunar orbit at a rate given by the coupling hypothesis ($\dot{r}_M/r_M \simeq H \simeq 2.5 \times 10^{-18}$ sec^{-1}) but with G constant. This means that we avoid the need to introduce G-variability provided we employ a gauge theory of gravitation with a space-time coupling of the expansion to small-scale systems. While this was explicitly realized by Wesson (1978a), it can be noted that the approximate coincidence between the rate of expansion predicted by the extended form of Hubble's Law and the observed recession speed of the Moon from the Earth had been remarked on earlier by King (1970). While the recession of the Moon from the Earth is affected by the dissipation of energy in the tides (Jeffreys, 1970) and possibly also by flexing of ice shelves (Doake, 1978), the coincidence noted by King still holds to within a factor of two. The newer result of Van Flandern (1975), which is based on the variation of the dynamics of the lunar orbit with atomic time and does not include a tidal friction term, puts the coincidence on a more exact basis. Whether or not the effect involved really is a consequence of space-time coupling in a gauge theory of gravity remains to be seen, but in principle this could be the case.

(b) The Hubble formula $\dot{r} = Hr$ can be combined with the expression for the mass-loss rate from an object of size r and with pressure p ($\dot{m} = -4\pi p r^2 \dot{r}/c^2$) to predict the rate of loss of mass by astronomical bodies. The Sun is a good example to take since we know what the mass-loss rate is by observations of the solar wind, so theory and observation can be compared immediately (Wesson, 1973, 1978a). The predicted mass-loss rate on the expansion hypothesis is $|\dot{m}| \approx 10^{12}$ gm sec^{-1}. The observed rate is somewhat uncertain but lies in the range $|\dot{m}| \approx 10^{11}$–$10^{13}$ gm sec^{-1}.

(c) As noted elsewhere (Chapters 5 and 10), there is some evidence, based on a range of geophysical considerations, which indicates that the Earth may be expanding at a rate of about 0.48 mm yr^{-1} (Wesson, 1973, 1975a, 1978a, b). The hypothesis of an Earth expansion that is coupled to the Hubble expansion predicts a rate of $\dot{r}_E = Hr_E \approx 0.51$ mm yr^{-1} with the value of H noted above.

The preceding three comments should be viewed as clues to a possible gauge theory of gravity and not necessarily as a proof that there exist processes in astrophysics and geophysics that cannot be explained by conventional theory. Each of the three coincidences (a), (b), (c) might be explained satisfactorily in itself by some special process to do respectively with the dynamics of the lunar orbit, solar physics and solid-Earth geophysics. The object of collecting together these coincidences is to show that in combination the three effects *can*, if it is desired, be interpreted in terms of a hypothesis that is consistent with a possible scale-invariant theory of gravity. Effects of a possible expansion of the type being considered are difficult to detect in other astrophysical systems. For a length $r \approx 10$ kpc, for example, the expansion velocity is $\dot{r} \approx Hr \approx 0.75$ km sec^{-1} ($H = 75$ km sec^{-1} Mpc^{-1}), and one might expect that the Milky Way would be expanding at about this rate on the basis of the coupling hypothesis. Ovenden and Byl (1976) studied the velocities of about 10^3 O and B stars, Cepheids and open clusters with a view to detecting a possible systematic expansion of the Galaxy. They did not detect any such process, but there are uncertainties in the analysis (e.g., the influence of the so-called K-term) that in effect restrict its accuracy to velocities of order 1 km sec^{-1}. Thus, the work of Ovenden and Byl (1976) does not rule out the expansion hypothesis at the level at which it is being discussed here. The authors note that their results are in fact compatible with Van Flandern's result for the variation of G and its possible dynamical consequences for stars in the Galaxy. There is no definite evidence for or against the coupled expansion hypothesis in astrophysical systems of other types. If one wishes to look for possible indications, other than (a)–(c) above, of gauge invariance in astronomy, one must consider systems which have the properties of being dominated by gravity and of being well enough studied that data are available with which to make a meaningful study. The Solar System satisfies these two criteria, and it is interesting to ask if there is any evidence of scale invariance in the planetary system and in the satellite systems of the planets.

An analysis that effectively shows that the bodies of the Solar System are arranged in a roughly scale-invariant manner has been given by Wesson (1979d). Actually, because the planets and satellites as they are seen today are discrete bodies, it is only possible to state that the (in practice) step-like distribution of the masses of secondary bodies around a given primary is close to the (theoretical) smooth distribution of a hypothetical self-similar disk of matter with the same total mass and angular momentum. This qualified statement of the self-similar nature of the matter distribution of bodies in the Solar System holds for the system Sun/terrestrial planets and for the planet/satellite systems of Jupiter, Saturn and Uranus. Suitable data are not available with which to test the self-similar nature of the mass distributions of other systems (for reasons to be given below). It must also be realized that although the noted systems are to a first approximation self-similar, the question of whether the Solar System is arranged in a scale-invariant manner really needs to be considered in conjunction with a theory of the origin of planets and satellites, since what one is really interested in is the nature of the continuous media (i.e., the disks) from which the present discrete systems originated.

A theory of the origin of planets and satellites that is compatible with the self-similarity revealed by Wesson (1979d) has been given by Wesson and Lermann (1978b) and Wesson (1978f). These three references together represent a self-similar theory of the origin of the Solar System. The question of whether the planetary system originated from scale-invariant matter distributions, in the forms of a disk around the Sun and disks around the proto-planets, depends on how well this theory of the origin of the planets and satellites fares in comparison with other theories which have been proposed and which are also in reasonable agreement with the observed Solar System.

The main conclusion of Wesson and Lermann (1978b) was that the dust from which the planets and satellites formed was charged by plasma processes like the dust in interstellar space, and that under such conditions accretion in the terrestrial zone occurred in two stages: firstly, electrostatic forces caused a separation of the material into metal (i.e., mainly iron) planetesimals of up to 10^2 cm size and a cloud composed of dielectric (i.e., mainly silicate) material in the form of small ($\approx 10^{-5}$ cm) grains; secondly, gravitational forces caused the metal planetesimals to clump together into bodies of the mass of planetary cores, which were later mantled by the silicate fraction. The time scale of the charged-dust accretion process would have depended on the time-evolution of the charge distribution among the original dust grains (Simons, 1976a, b), and on the fctor $(\epsilon - 1)/(\epsilon + 1)$ which regulates the difference in electrostatic accretion force for grains composed of material with different dielectric constants ϵ (Simpson, 1978). A simple model (Wesson and Lermann, 1978b) leads one to expect that the iron planetesimals formed in a time of the order of 10^7 yr, clumped in a time of 10^5–10^7 yr, and that the silicate fraction was captured in a time not exceeding 10^8 yr.

The formation of planets with iron cores was studied further by Wesson (1978f), who also reviewed mechanisms that have been proposed to account for the formation of iron bodies in the early Solar System. A fractionation mechanism of this type is required in order to explain a range of chemical and physical data to do with the planets and meteorites (see Lewis, 1972a, 1973; Delsemme, 1977; Wesson, 1978f; Sears, 1978). In particular, 'left-overs' from the grain/grain accretion and planetesimal formation processes can be identified respectively with the cosmic spherules of size of order 10^{-2} cm found today in deep-sea clays (Parkin, 1978) and with the meteorites. Members of both these classes of object are roughly classifiable as irons and stones, supporting the idea of an accretion process of the inhomogeneous type. In addition to the mechanisms mentioned in Wesson (1978f), three other processes have been proposed recently that might also have caused a separation of metallic and non-metallic material in the early Solar System: (1) Gladyshev (1978) has argued that chemical reactions in the solar nebula would have caused iron to separate from the gas first and form metallic bodies; (2) Weidenschilling (1978) has suggested that the dependency of the rate of decay of planetesimal orbits by gas drag on the sizes and densities of the planetesimals concerned could have been responsible for iron-silicate fractionation in the early Solar System, and that such a process is compatible with the presence of a small dense planet like Mercury in the inner part of the System; (3) Slattery (1978; see also Slattery *et al.*, 1978) has argued that a rain-out of liquid iron in gaseous protoplanets would have produced iron cores at their centres. All three of these accounts can be employed to explain how material of metallic and rocky type became separated in the early Solar System, but they appear to be less likely than the charged-dust mechanism in terms of the conditions required for their operation.

A model for the early history of the Earth has been discussed by Clark *et al.* (1972) in terms of the formation of a metallic core from Fe-Ni condensates in the solar nebula. This model is based on work of Grossman (1972) which deals with equilibrium condensation in the primitive nebula; and (4) the condensation of Fe-Ni alloys at higher temperatures than other substances represents the main viable alternative to the charged-dust process for the formation of proto-planetary metallic cores. The early chemical history of the Solar System has been discussed by Grossman and Larimer (1974; see also Ward, 1976; Lewis, 1973). There is good evidence from cosmochemical data that iron bodies did indeed form first in the early solar nebula, even though these data do not make it clear just what formation process was involved.

The inhomogeneous accretion hypothesis is also in agreement with isotopic abundance data, which indicate that the Earth's core formed early on in the history of the planet (Oversby and Ringwood, 1971; see also Shaw, 1978; Jacobs, 1970). It must be admitted though that such data can be interpreted in different ways; and while the core seems to have been formed early on (i.e., within 10^7–10^8 yr of the origin of the planet), this does not by itself necessarily imply that it originated in an inhomogeneous accretion process (Vollmer, 1977). While some data exist which suggest that the Earth has always possessed an iron core (Wesson, 1978a, f) – including the fact that the geomagnetic field existed 3×10^9 yr ago (Nagata, 1970) – the time scale of its 'formation' with respect to the planet as a whole would have depended considerably on the nature of the accretion process that formed the planet. If the charged-dust mechanism was responsible, an iron core already existed before the mantle (i.e., the rest of the planet) was accreted. Reorganization may have occurred depending on how much silicate was mixed up with the iron core (some silicate must certainly have been mixed into the core, since the iron-silicate separation mechanism would not have been perfectly efficient). Reorganization may have altered somewhat the form of the core after the complete planet was formed, but basically the 'formation' time scale, as far as it can be defined on this model, is the silicate accretion time scale (i.e., 10^7–10^8 yr). If the alternative condensation mechanism (4) was responsible, the time scales of core and mantle formation would have depended more directly on the temperature and pressure in the primeval nebula (see Lewis, 1972a, b, for temperature-dependent fractionation processes). The process of condensing out of the Fe and Ni would probably have only taken of the order of 10^4 yr (Clark *et al.*, 1972), and depending on conditions in the nebula the whole planet would certainly have been complete in 10^7–10^8 yr, and possibly in a much shorter time. Thus, for both the charged-dust and condensation theories, the 'core formation time', judged with respect to the planet as a whole, was less than or of the order of 10^8 yr, in agreement with the geophysical data noted above.

The formation of planets with iron cores is a physical process which, by itself, might also be compatible with theories of the origin of the Solar System that do not make use of self-similarity. Cases in point are those theories of the origin of the Solar System that ascribe importance to forces other than gravity in the formation of the planetary system and the satellite systems of the planets. For example, there is evidence that magnetic fields with strengths of up to 16 Oe existed at the time the chondrules which make up many meteorites were formed (Strangway, 1978; see also Levy, 1978); and in the present-day Solar System several of the planets (Mercury, Earth, Jupiter, Saturn and Uranus) have finite magnetic moments (Ness, 1978). The problem of the transfer of angular momentum from the Sun to the planets has also been studied in relation to a proto-solar magnetic field that might have dominated the motion of matter in an ionized solar nebula. There are several variations of this mechanism for transferring angular momentum from the Sun to the planets, and altogether there are almost as many proposed transfer mechanisms as there are theories of the origin of the planets; see Freeman (1978) for the strength of the primordial solar magnetic field, and Cameron (1963), Woolfson (1969), Ward (1976) and Dermott (1978) for reviews of the origin of the Solar System.) However, in general, the operation of forces to do with magnetic fields would *not* have resulted in the presence of disks with self-similar density laws in the early Solar System. What is quite definite is that the mass distribution of the terrestrial planets and the satellite systems of Jupiter, Saturn and Uranus are compatible with the origin of these systems from *gravitationally* dominated disks with self-similar density laws (Wesson, 1979d). This shows that if non-gravitational forces did at any time play an important role in the dynamics of the material of the solar nebula, then it was at epochs prior to that at which the planets and satellites fragmented out of their parent disks. In view of this, it seems natural to account for the formation of planets and satellites by a two-component theory in which electrostatic grain/grain forces on small scales (10^{-5}–10^2 cm) caused a separation of material into metals and silicates (Wesson and Lermann, 1978b; Wesson, 1978f), and in which gravitational forces on large scales (10^5–10^8 km) arranged the material concerned into self-similar disks and led to its eventual accumulation into bodies of planetary mass.

The combined theory of the formation of planets outlined above starts from the existence of a dust disk around the Sun and ends with the existence of bodies possessing iron cores and silicate mantles arranged in an approximately scale-free distribution similar to that which is seen today. As such the theory encompasses the first three of what Kaula (1975) has called the seven ages of a planet: condensation, planetesimal interaction, formation, convection, plate tectonics, terminal volcanism and quiescence. The last four stages concern what happens to the planet after it first comes into existence as a discrete body, and while of no direct relation to the self-similar theory of planetary formation, some of the processes which a planet might be subject to after its formation do have relevance to the comments of Chapters 2 and 5.

The infall of the planetesimals left over from the formation process would probably have heated up a planet like the Earth considerably, perhaps melting its outer layers (Safranov, 1978). Such heating of the Earth and the other terrestrial planets might have resulted in further differentiation of previously mixed material inside the protoplanets (Toksöz et al., 1978; Toksöz and Hsui, 1978). Afterwards, their mantles would have settled down into the thermal regimes which are inferred today from studies of planetary morphology (Runcorn, 1967; Windley, 1976). Planetary atmospheres probably formed by the release of gas from material that became heated during the later stages of planetary formation (see, e.g., Sill and Wilkening, 1978). Thereafter, the atmosphere of a planet like the Earth would have evolved in accordance with a variety of things such as the size of the solar constant (i.e., the amount of heat reaching the Earth from the Sun), the planet's albedo and the operation of a possible greenhouse effect due to the cloud cover (Sagan and Mullen, 1972; Sagan, 1977; Henderson-Sellers and Meadows, 1977). A consideration of these things leads one to expect that the Earth's surface was warmer 3×10^9 yr ago than it is now (Henderson-Sellers and Meadows, 1977). This inference is supported by a study of theoretical temperature profiles for a chemically-evolving atmosphere, which implies that the Earth's paleoatmosphere was about 10°C warmer for epochs prior to 10^9 yr ago than it is now (Morss and Kuhn, 1978). These paleoclimatic models are confirmed in their main contention (viz: that the temperature T_E of the Earth was higher in the past) by data on hydrogen and oxygen isotope ratios for ancient cherts (Knauth and Epstein, 1976). These latter data indicate that T_E has decreased through geological time, possibly having been as high as $T_E \simeq 52°C$ at 1.3×10^9 yr and $T_E \simeq 70°C$ at 3×10^9 yr in the PreCambrian.

The main influence on the Earth's surface temperature is the solar constant, and conditions are so sensitive to this parameter that life can only comfortably exist on the Earth given its present type of atmosphere for a very narrow range of 0.95–1.01 AU for the orbital distance of the planet from the Sun (Hart, 1978). This sensitivity raises a problem, since most conventional stellar evolution calculations indicate that the Sun's luminosity has increased by about 25% since its formation. This implies that the ancient Earth should have been notably cooler than it is now (Newman and Rood, 1978), contrary to evidence indicating that it was in fact warmer (Knauth and Epstein, 1976). This problem might be solved in terms of a cosmology in which G was larger in the past than it is now (see Chapter 2). However, an equally plausible explanation is that the greenhouse effect and other effects to do with the structure of the Earth's past atmosphere have compensated for any secular change in the solar constant.

It is not easy to separate out possible cosmological-based effects from astrophysical and geophysical ones where the past nature of the Earth's atmosphere is concerned. Even the survival of a planetary atmosphere depends on astrophysical processes to do with the past behaviour of the Sun, such as the intensity and frequency of solar storms and the density of the solar wind. (Some evidence for a vanished lunar atmosphere has been discussed by Chernyak (1978) and commented on by Hughes (1978), but the question of whether planetary bodies which are now airless or nearly airless ever had more substantial atmospheres remains open to debate.) Geophysical effects are probably less important (e.g., Miles and Gildersleeves (1978) have shown that the presence of dust in the atmosphere from volcanic eruptions is not correlated with present-day changes in global temperature); but they cannot be ruled out as insignificant in view of uncertainties about the thermal regime of the upper mantle and lithosphere in the past (see, e.g., Burke and Kidd, 1978a, b; Drury, 1978). Furthermore, the surface temperature in the past of a planet – like the Earth – which possesses a substantial atmosphere has been seen to be dependent to a certain extent on purely meteorological factors, the influence of which are difficult to assess accurately. One is left with the impression that uncertainties such as those noted here make the estimation of the Earth's past temperature a very risky business. It would certainly seem that attempts to evaluate the expected effects of variable-G cosmology on the temperature of the Earth (Chapter 2) are unlikely to come to any definite conclusion.

This comment should not be regarded as an attempt at discouragement but rather as a caution against using data from geophysics and planetary physics as too-definite evidence either for or against cosmologies of the type discussed in Chapter 2. In fact, it is clear from the discussions of Chapters 2 and 5 and the preceding paragraph of the present chapter that a profitable line of research in this area would be to construct a (possibly computerized) model of a planet which is affected by G-variability, expansion and the geophysical effects (such as convection and atmospheric evolution) which one might expect to be important on the grounds of purely Newtonian physics. One could in this way work out the evolutionary history of a planet like the Earth as it is influenced by the noted factors. A comparison of the model with the actual Earth would then show how the different processes interrelate, and perhaps allow limits to be obtained on some of them.

On the scale of the Solar System, another interesting question which one can pose concerns whether a planetary system which is affected by G-variability or expansion evolves in a self-similar manner. That is, does the density run remain self-similar as the system changes its dimensions? Those parts of the Solar System which have homogeneous chemical properties and are well studied (i.e., the terrestrial planets and the satellite systems of Jupiter, Saturn and Uranus) are, as discussed above, arranged in a roughly scale-invariant manner (Wesson, 1979d). If G is changing with time, or if these systems are expanding, the self-similarity observed at the present epoch cannot be an accident: it has to be conformable to the G-variability or expansion process involved. By turning the problem around, it may be possible to obtain restrictions on the forms of these processes by imposing the condition that they should be conformable with the preservation of scale invariance.

Further work on the topic of scale invariance in the Solar System might be aimed at refining the self-similar theory of the origin of the planets. This theory is already in a strong position as regards comparison with other theories of the origin of the Solar System (Jastrow and Cameron, 1963; Woolfson, 1969; Ward, 1976; Dermott, 1978), for the following reason: The self-similar theory fits the observations for the four systems noted above *without* bringing in adjustable dimensional parameters. The absence of adjustable dimensional parameters, which is a consequence of the self-similar requirement, makes the theory very vulnerable to disproof and therefore logically well founded. Put bluntly: the self-similar theory has to be either right or wrong, and cannot be 'adjusted' into agreement with the data like most other theories of the origin of the Solar System. However, the self-similar theory has its restrictions. The major one is that it cannot be applied to the system of the giant planets as these latter are observed today because it is clear from their chemical compositions that these planets have lost mass in disproportionate amounts since the disk-phase of the Solar System, and without information on this the self-similar approach cannot be employed. (The same problem does not arise in the case of the terrestrial planets, since any mass loss that these planets have undergone seems to have been uniform, meaning that all planets lost the same proportion of mass, so the original self-similar structure of the disk was preserved.) It would be worthwhile to try to reconstruct the form of the disk from which the giant planets formed using cosmochemical data (see Ponnamperuma, 1976), and see if it agrees with the form predicted by the self-similar theory.

The type of scale invariance of astrophysical systems in general which is revealed by the successful agreement of self-similar models with observation is in the slightly unusual position of being established but not fully understood. That is, many astrophysical systems can be described successfully by self-similar models, but the basic meaning of this form of scale invariance has not been related to an underlying principle of the theory of gravitation. (By comparison, the relation of scale invariance to scalar-tensor theories of gravity has been studied in some detail, as described in Chapter 8; and scale-invariance has been taken as a foundation for some of the theories of gravity alternative to general relativity listed in Chapter 9. Scale invariance might also be understandable as a prerequisite for the successful application of group theory to gravity as discussed in Chapter 10.) The problem of finding a fundamental connection between self-similarity and gravitation theory represents another topic of worthwhile research. It may be that the connection can be understood without going outside the bounds of general relativity, or it may prove to be necessary to widen the theory to include a gauge-invariant description of gravitational phenomena as suggested above. The latter question is most likely to be decided in practice by experiment. In particular, long-term spin-down effects for rotating masses due to a coupled-expansion theory can probably be tested for directly by using an experiment of the type developed by the University of Virginia group (Wesson, 1978b). The gravity experiments of different types which have been discussed previously should also allow one to answer with certainty the question of whether there exist physical processes (such as a variation in G or a change in the masses of bodies) that operate on time scales longer than the age of the Universe and which are not included in Einstein's theory.

CHAPTER 7

SUMMARY

It is appropriate here to summarize the contents of the foregoing chapters and the inferences which can be drawn from the more technical material contained in the succeeding ones, before listing the main oustanding problems of the subject in its entirety that deserve further attention.

In Chapter 2, the three main variable-G theories were examined (2.2) and their status with respect to other theories (2.3) shows that they represent the best possibility for advancement in the field of gravitation, mainly because of their compatibility with the concept of scale invariance in one or another of its various forms. In Chapter 3, the connection between particle physics and gravitation was discussed, notably as such a connection might involve gauge theories of particle physics (3.2) and cosmology (3.3). While no firm connection yet exists between particle physics and gravitation, there are numerous signs that such a connection will be found, probably via the cosmological constant. In Chapter 4, the relationship of gravitation (and variable-G theories in particular) to astrophysics was examined. Astrophysical systems provide opportunities for testing G-variability, but so far the inferences which have been drawn from studies of such systems have been mainly negative as they involve a possible time-dependence of G, although the Dirac ($\times$)-model appears to be compatible with a wide range of astrophysical data. In Chapter 5, it was seen that geophysics may provide evidence in favour of a new theory of gravitation in the form of the expanding Earth hypothesis. This hypothesis is as yet unconfirmed, but the effect involved would seem to be too large to be explained by a simple time-dependence of G. In Chapter 6, a discussion of various aspects of gravitation and astrophysics showed that the best hope for progress in measuring G-variability and variability of the masses of bodies on secular time scales lies in laboratory experiments (6.2). These experiments will provide a direct test of scale-invariant gravitational theories such as that of Dirac, and also help to determine if there is evidence in favour of a gauge theory of gravity which is scale invariant in the sense of being an extension of self-similar general relativity (6.3). Chapter 8 discusses scale invariance and scalar-tensor theories, but these latter do not appear to be favoured by observational data. Chapter 9 treats theories of gravity that are alternative to general relativity, and contains a compendium of viable gravitational theories (9.2) and a discussion of how they compare one with another (9.3). Chapter 10 is concerned with group theory and gravity, in particular with the conformal group and related topics (10.2) and other aspects of group theory as it relates to gravitation (10.3). Chapter 11 discusses the status of non-Doppler redshifts in astrophysics. Various gravitational theories predict the existence of such redshifts, but a reveiw of the tired-light and related theories (11.2), redshift anomalies in astronomical sources (11.3) and non-Friedmannian redshifts (11.4) leads one to the conclusion that there is no firm evidence for the existence of redshifts produced by non-Einsteinian processes. However, there do appear to be departures from the isotropic Hubble law of redshifts, probably connected with departures of the real Universe from the idealized Friedmann models due to the clustering of galaxies.

Having completed the preceding survey, one might be allowed to give some opinions on the subject. There are two main comments that can be made, one of them negative and the other positive in tone.

The negative comment is that, despite the enormous amount of effort which has been spent on constructing non-Einsteinian theories of gravity and hypotheses that underpin the idea of G-variability, there is still no definite support for any of these new accounts. Indeed, there is a lot of evidence against theories of gravity other than the main G-variable ones of (a) Dirac, (b) Hoyle/Narlikar and (c) Canuto *et al.*; and even these theories face increasingly stringent limits on a possible time-variability of G, the only outstanding non-null result being that of Van Flandern (1975). Observational constraints must, of course, take priority over theoretical considerations, irrespective of how compelling the latter may be; and if G-variability proves to be observationally untenable, then it will have to be abandoned.

The positive comment is that there do exist indications that a new theory of gravity will some day supersede general relativity and widen the scope of research in gravitation and cosmology. (This comment and the discussion of the first part of the text in general refer to non-quantized theories of gravity. Einstein's theory will also be superceded if a way of quantizing the gravitational field is found. This possibility is mentioned in Section 10.3, where references to the subject are given.) The indications of a new classical theory of gravity come from various sources, and while maybe not significant individually, in combination do tend to suggest that a viable gauge theory of gravitation exists. The clues involved concern among other things the various tests which have been applied to the Dirac ($\times$)-theory. As seen in Chapter 2, this form of the Dirac theory shows a surprising ability to survive comparison with different types of observational test. This staying-power cannot be accidental. But neither does it necessarily prove that the Dirac theory is correct, since the Large Numbers Hypothesis on which it is based may be compatible with other theories of gravity. Further clues to a possible new theory concern the expanding Earth hypothesis (which indicates an effect requiring more than G-variability for its explanation) and the compelling nature, both theoretical and observational, of various aspects of the concept of self-similarity. Since progress cannot be stopped, one can be excused for speculating that perhaps it will in the future prove possible to understand these disparate pieces of information as attributes of a single fundamental theory.

Meanwhile, there are some problems that can profitably be investigated with a view to clarifying what form a possible new theory of gravitation might take. The main outstanding points are the following:

(a) Is G varying with time, and if so, how fast? This question will hopefully be answered soon by laboratory experiments, at least as regards a variation at a rate of order $|\dot{G}/G| \approx 10^{-11}$–$10^{-12}$ yr^{-1}. But the related problem of a secular change in the masses of all bodies is equally important, and there are grounds for believing that the rate of such a process would only be $|\dot{m}/m| \approx (p/\rho c^2)H$ where p is the pressure and ρ is the density of the body concerned (Wesson, 1978a). Even in the interior of the Sun, $p/\rho c^2 \approx 10^{-4}$, and as regards geophysical and laboratory tests the factor $p/\rho c^2$ would be still smaller. Even if an experiment could be devised which achieved a pressure of the order of that in the Sun's interior, the mass-loss rate with a value of Hubble's parameter $H \approx 10^{-10}$ yr^{-1} would still be only $|\dot{m}/m| \approx 10^{-14}$ yr^{-1}. It would take a very sensitive measuring device to detect directly a mass change at a rate of this order in the laboratory, but it may be possible to test the effect in other ways.

(b) Is there a connection between particle physics and gravitation, and if so, what form does it take? It has been seen above that the first part of this question will in all probability be answered in the affirmative as regards the weak interaction and electromagnetism, but the possibility of bringing the strong interaction into the scheme is still speculative.

(c) Is there a formalism in which gauge invariance in the sense of the Weyl geometry for cosmology can be identified with the gauge invariance of theories of particle physics? The two concepts have a common conceptual meaning, but mathematically it is not clear if the ways in which they are used in gravitation and particle physics are equivalent. There is, indeed, some confusion as to what different workers mean by terms composed of combinations of the four words gauge/scale/invariance/covariance (and, in cosmology, of the terms conformal invariance and self-similarity also; see Chapter 6 for possible research topics involving self-similarity). A useful project would be to examine the mathematical content of the various ways in which these terms are used in both gravitation and particle physics, and arrange them in a hierarchy, ordered according to the generality and power of the terms concerned.

(d) What kind of cosmological model should one use in order to evaluate the astrophysical and cosmological consequences of new theories of gravity? In the past, most workers have assumed the form of the Robertson/Walker line element and used this in conjunction with a given theory to evaluate astronomical consequences of that theory which could then be compared with observation. Sometimes the predictions and the observations did not agree, and in this case the prevalent opinion seems to have been that the theory concerned was wrong. Another possibility, of course, is that the cosmological model used in deriving the predictions was wrong. In recent years there has been a considerable upswing in interest in non-Friedmannian models of the Universe, with line elements different to that of the Robertson/Walker line element. The main reason for this has been a growing awareness of the fact that information about the clumping of galaxies on large scales is really quite uncertain: it is not obvious that the Universe actually is isotropic enough to justify the use of the Friedmann models. It would therefore appear to be judicious to examine the status of new theories of gravity with a range of cosmological models, before deciding whether or not a given theory is viable.

(e) Is the Earth expanding, and if so, at what rate? This question has two aspects to it, both of which are of considerable significance for gravitational theory. Firstly, if G is decreasing on a time scale of order $H^{-1} \approx 10^{10}$–10^{11} yr, then the Earth should be expanding at a rate of order 0.02–0.3 mm yr^{-1}. Secondly, there is already a body of data which suggests expansion, but at a rate of about 0.5 mm yr^{-1}, which may be indicative of a new theory of gravity with more drastic consequences than those theories which incorporate the $G \propto t^{-1}$ hypothesis in its usual form. What geophysical evidence there is on expansion is disputed. New, reliable data on a possible expansion of the Earth would be very useful in helping to clarify matters. Also, a computer model of the structure of a planet like the Earth which is affected by G-variability and expansion would be a useful tool for use in evaluating the geophysical consequences of these processes.

An answer to, or elucidation of, any one of the five questions listed above would represent a significant contribution to research in the theory of gravitation. Several of the listed problems can best be tackled by multi-disciplinary approaches; but this should not discourage individual attempts at their solution, since in the nature of the subject there is a wide scope here for the practice of ingenuity. The major motivation behind following the paths for future research represented by the questions (a)–(e) is that an answer to any of them would provide a clear pointer to a theory of gravitation to succeed general relativity.

PART TWO

CHAPTER 8

SCALE INVARIANCE AND SCALAR-TENSOR THEORIES

In this chapter, a brief discussion is given of gauge or scale invariance as it relates to gravitational theories with an added scalar field. (Scale invariance in a wider context is also discussed in Chapter 9.) The following three paragraphs comment in this respect on scalar modifications of general relativity, the Brans/Dicke theory, and scalar-tensor theories of wider scope that were inspired by the Brans/Dicke theory.

An attempt by Callan et al. (1970) to make Einstein's theory scale invariant consisted in introducing a new, improved energy-momentum tensor T_{ij}. This innovation, which was inspired by quantum mechanics, involves the addition of a scalar field to gravity that renders T_{ij} scale invariant. A similar approach was made by Deser (1970), who employed scale invariance in the form in which it means that physical systems are describable in terms of dimensionless parameters, to construct a new action from which a theory can be derived which is essentially a variant of general relativity but with G variable. (An earlier attempt by Gilbert (1956, 1960) at incorporating G-variability into general relativity has been discussed by Wesson (1978a). It was essentially a predecessor of conformally invariant theories like that of Hoyle/Narlikar.) Deser, like Callan et al., introduced a scalar field which can account for the breaking of scale invariance in some areas of physics if it has a finite range of interaction with matter. (In practice, the horizon of the visible Universe, defined by $r \equiv ct_0 \simeq cH_0^{-1} \simeq c \times 10^{10}$ yr where H_0 is the present value of Hubble's parameter, is a typical length scale in cosmology provided one regards $t_0 \simeq 10^{10}$ yr as a characteristic time scale for large-scale astrophysical processes. Alternatively, the length $c\Lambda^{-1/2}$, where the cosmological constant is taken with the dimension time^{-2}, is a possible cosmological scale that can provide a range restriction for any scalar field of cosmological type. An axiomatic renormalization of the energy-momentum tensor for a conformally invariant field has been considered by Wald (1978) for the case of conformally flat space-times.) The departures of Callan et al. (1970) and Deser (1970) yield theories which can be made to satisfy the classical tests of general relativity at *some* level, and are similar to the Brans/Dicke theory (Brans and Dicke, 1961) insofar as all three theories possess a scalar interaction in addition to the usual tensor interaction of unmodified general relativity.

Invariance under arbitrary space-time dependent changes of scale is consistent with Brans/Dicke theory only for the special choice $\omega = -3/2$ for the coupling parameter (Anderson, 1971). Unfortunately, for this choice of ω the scalar field which lies at the heart of the theory can be transformed such that it becomes a constant everywhere, and the Brans/Dicke theory in this instance is equivalent to Einstein's general relativity. (As discussed by Thorne and Dykla (1971) and Bhamra *et al.* (1978) the structures of black hole and naked singularities in the two theories are similar for all values of ω.) The Brans/Dicke theory also tends to general relativity for $\omega \to \infty$, in which limit $\dot{G} \simeq 0$ and the theory becomes indistinguishable from Einstein's in regard to both the classical Solar System gravity experiments and cosmology. The fact that the Universe on both these scales appears to be in approximate agreement with general relativity has led Barker (1978) to suggest that Brans/Dicke theory may nevertheless be valid, but in a kind of impotent form in which ω changes with time and just happens to be large at the present epoch. Barker's hypothesis involves ω changing from $\omega = -3/2$ (in the distant past), passing through $\omega = \infty$ (now), going on to $\omega = 0$ (maximum extension of the Universe), and then back to $\omega = -3/2$ (in the distant future). This scenario can be accomplished in a finite time, and if true would lead to a recollapse of the Universe in the future. (Ruban (1977) has discussed the effect of a scalar field on the structure of cosmological singularities.) However, although Barker's suggestion is formally acceptable in making the rate of change of the gravitational parameter very small at the present epoch and in avoiding conflicts with other results that are in agreement with general relativity (see below), it sounds very *ad hoc*. On the more positive side, the Brans/Dicke theory, both in its original ($\omega \simeq 6$, $\dot{G}/G \simeq -2 \times 10^{-11}$ yr^{-1}) and more contrived forms, does possess some cosmological-type solutions of reasonable type (Caloi and Firmani, 1971; Morganstern, 1971a, b, c, 1972; Dehner and Obregón, 1971, 1972; Matzner *et al.*, 1973; Lessner, 1974; Occhionero and Vagnetti, 1975; Bishop, 1976; Barker, 1978; Obregón and Chauvet, 1978). It is possible to choose parameters so that the theory is compatible with data on the mean density and age of the Universe, Hubble's parameter H_0, the deceleration parameter q_0 and the magnitude/redshift (m/z) diagram for galaxies (Barnes and Prondzinski, 1972; Morganstern, 1973). It also predicts cosmic abundances of D, He^3 and He^4 that are about the same as those given by general relativity if the cosmological model concerned is an open one, although too much He^4 is predicted by the theory if the cosmological model is a closed one (Greenstein, 1968). But the Brans/Dicke theory is in conflict with Solar System data for any value of ω near to the value $\omega \simeq 6$ first advocated by Dicke. Laser tracking of the

Moon's orbit has enabled Williams *et al*. (1976) to set the limit $\omega \gtrsim 29$ for the coupling parameter, a result which has been essentially confirmed by Shapiro *et al*. (1976). This limit, and others of comparable sizes (Wesson, 1978a), means that the scalar-tensor theory is effectively only acceptable in the quasi-Einstein regime of large ω. As noted above, one can in principle preserve the Brans/Dicke theory by postulating special values and behaviours for ω, but this is difficult to justify. For example, it has been suggested by Smalley and Eby (1976) that if ω is negative then the theory can be brought into agreement with data concerning the deflection of light by the Sun and limits on $\dot{G}/G$. However, while the theory might formally be reconciled with experiment by adopting the assumption that ω can be negative along with Smalley and Eby (1976) and Barker (1978), there are good reasons, based on the Principle of Least Action, for believing that the coupling parameter in the Brans/Dicke theory has to be positive (Noerdlinger, 1968). As far as experiment and observation go, it must also be stated that, irrespective of the sign of the coupling parameter, there do not appear to exist any anomalous astrophysical data that justify the introduction of a scalar-tensor theory in preference to general relativity. In particular, while it used to be possible to make a case that the shape of the Sun showed a certain departure from Newtonian theory that could have been indicative of a scalar interaction in gravity (see, e.g., Dicke, 1970), more recent data (summarized in Wesson, 1978a) show that there does not exist any significant anomaly that cannot be accounted for in terms of conventional solar processes (Schatten, 1975, 1977). Thus, although the Brans/Dicke theory has been investigated in considerable detail as regards its agreement (or otherwise) with the classical Solar System tests of general relativity (Wagoner, 1970; Nordtvedt, 1970b; Morganstern, 1971d; Ni, 1972; Lee, 1974), as stated in Chapter 2 of the text it is physically only acceptable in forms in which its consequences differ but insignificantly from Einstein's theory.

The accounts of Bergmann (1968) and Wagoner (1970) both deal with scalar-tensor theories that were inspired by, but are of wider scope than, the Brans/Dicke theory. Bergmann has discussed the effects of a scalar field on the momenta and masses of particles in a generalized scalar-tensor theory; while Wagoner has discussed how such theories compare with the results of the classical tests of Einstein's theory, and has commented on the generation of gravitational waves in such theories. The Brans/Dicke theory is a special case of the Bergmann/Wagoner theory. (The theory of Jordan (1949, 1962), which has been discussed by Wesson (1978a), was a forerunner of the theory of Brans and Dicke (1961), and the complete account is often referred to as the Jordan/Brans/Dicke theory.) The Nordtvedt scalar-tensor theory

(Nordtvedt, 1970b; Ni, 1972) and the Bicknell/Klotz theory are also special cases of the Bergmann/Wagoner theory. The Bicknell/Klotz theory is worth commenting on since it was directly based on the concept of conformal invariance. After previous work on a geometrical Lagrangian for the neutral scalar meson field (Bicknell, 1974a) and the realization that gravitational theories based on a quadratic Lagrangian (i.e., ones depending on the square of the scalar curvature of space-time) conflict with observation even if they are mathematically consistent (Bicknell, 1974b), Bicknell (1976) examined the status of conformal invariance in a theory consisting of general relativity plus a scalar field ϕ. He confirmed that particle masses have to vary as $m = m_0\phi$ (where m_0 is a constant) if conformal invariance (i.e., invariance under gauge transformations of the type $g_{ij} \rightarrow \beta^2(x)g_{ij}$) is to be preserved in a theory consisting of general relativity plus a massive spinor field introduced from particle physics. This subject had been earlier discussed by Bramson (1974), who was interested in how one's interpretation of particle mass depends on the choice of one's gauge, and who showed that general relativity can be derived from a conformally invariant theory of the gravitational interaction provided one makes a special choice of gauge (in which G and particle masses always occur together). The field equations of a conformally invariant scalar-tensor theory were derived from an action by Bicknell and Klotz (1976a). Their theory differs slightly from the Brans/Dicke theory, having only moderate G-variability and a term which corresponds to a variable Λ. It is a special case of the Bergmann/Wagoner theory, characterized by having reasonable cosmological solutions (Bicknell and Klotz, 1976b) and, relative to the Brans/Dicke theory, a quite strong interaction with matter on the Solar System scale. But while the Bergmann/Wagoner theory and its special cases may be acceptable at some level as descriptions of gravity (Wagoner, 1973), one must note that the existence of such theories does not by itself provide justification for a departure from general relativity.

CHAPTER 9

ALTERNATIVE THEORIES OF GRAVITY

9.1. Introduction

New accounts of gravity are usually inspired by some theoretical consideration which is taken as a basis for the further development of the theory. The degree to which that development is carried out varies greatly, and it is often the case that a new gravitational theory is worked out in considerable detail in some areas and in minimal detail in others. This concerns especially observational consequences, which are well investigated in some theories and poorly investigated in others.

The main part of the present chapter represents a compendium of theories of gravity that may be compatible with observation and which are alternative in the sense of being non-identical to the standard theory of Einstein's general relativity.

9.2. A Compendium of Gravitational Theories

The thorough examination of metric theories of gravity by Ni (1972) mentioned in Chapter 8 showed that, of theories which had been fully developed up to about that time, only the reference theory (0) of general relativity (possibly modified as discussed in the first part of Chapter 8) and (1) the Bergmann/Wagoner theory (with its special cases of the (1a) Jordan/Brans/Dicke, (1b) Nordtvedt and (1c) Bicknell/Klotz scalar-tensor theories) were in acceptable agreement with observation. To this list one must add the main variable-G theories discussed in Chapter 2 of the text, namely those of (2a) Dirac, (2b) Hoyle/Narlikar and (2c) Canuto *et al.* (3) The vector-tensor theory (Hellings and Nordtvedt, 1973), which has been discussed elsewhere (Ni, 1972; Wesson, 1978a), is also acceptable.

Of other theories of gravity, one class that can be conveniently discussed includes those theories which attach special significance to a frame or gauge in which space-time is flat.

(4a) Prokhovnik (1970a) proposed a cosmological theory of gravitation which is based on the assumption that Hubble's law holds in the form $\dot{r} = r/t$ with $H = 1/t$ in a flat ($k = 0$) Robertson/Walker space-time. The flat space-time defines a basic reference frame for the propagation of light, and the theory differs from Newtonian mechanics and from general relativity in that a particle which moves with a given velocity past an observer who is attached to the substratum proceeds to pass all similar observers with the same velocity. Newton's first law of motion now only holds exactly for bodies at rest in the substratum. This model is consistent with the principles on which Einstein's theory of relativity is based even if it does not use field equations as in general relativity (Prokhovnik 1964, 1967, 1968, 1970a, b). It is also consistent with modern cosmology in that the galaxies are assumed to form a class of particles that recede from each other as a result of a uniform expansion of the Universe (Prokhovnik, 1970a, b). The galaxies have redshifts z that are determined by their motions in combination with a theory of light propagation modified in accordance with the condition noted above, this yielding redshifts that are linearly related to the distance from source to observer for small values of z (Prokhovnik, 1964, 1968). The noted condition on the velocity of a particle, which moves partly because of the expansion of the Universe and partly because of its intrinsic peculiar velocity, results in an equation for the acceleration which differs from the conventional one, being now

$$\ddot{r} = -\frac{r}{t^2} + \frac{\dot{r}}{t}. \tag{9.1}$$

This equation is similar to that of (2.4) of the first part of the text in that the first term can be identified with a G-dependent gravitational force (see the next paragraph) while the second term is essentially cosmological, and is just $H\dot{r}$. The total acceleration $\ddot{r}$ thus consists of two parts, the first of which $(-r/t^2)$ is small while the second $(\dot{r}/t)$ is large.

The first term represents a decrease in the velocity $\dot{r}$ for a particle at some fixed value of r (i.e., the contribution to $\ddot{r}$ from this part is negative and depends inversely on t for fixed r). Prokhovnik assumed that this part could be rewritten as $-(r/t^2)(\rho/\rho_U)$ where ρ is the local value of the density of matter at any place in a Universe whose mean density is ρ_U (taken to be $\rho_U = 2 \times 10^{-29}$ gm cm^{-3} with H at the present epoch given by $H_0^{-1} = t_0 = 4 \times 10^{17}$ sec $\simeq 1 \times 10^{10}$ yr). Writing $\rho = 3M/4\pi r^3$ for the local density, the part of the acceleration due to the first term in (9.1) can be expressed as $-GM/r_2^2$ where $G = 3/4\pi t_0^2 \rho_U \simeq 7 \times 10^{-8}$ gm^{-1} cm^3 sec^{-2} as measured in the laboratory when one substitutes the noted values of ρ_U and t_0. This type of relation between G and t_0 is common to many accounts of inertia and gravitation (Wesson, 1978a). But one should notice that it only works numerically because the value of ρ_U assumed ($\rho_U \simeq 2 \times 10^{-29}$ gm cm^{-3}) is considerably larger than the value arrived at from observations of the luminous matter in the galaxies, which give $\rho_U \simeq 2 \times 10^{-31}$ gm cm^{-3}

(Shapiro, 1971). Most arguments for taking the larger value of ρ_U are unsatisfactory (e.g., $GM_U/c^2R_U \simeq 1$ as in Equation (10.6) of the next chapter for the 'mass' of the Universe if ρ_U is of order 10^{-29} gm cm^{-3}, the presence of unity on the right-hand side of this last relation being aesthetically more pleasing to some than a value of order 10^{-2}). However, the presence of dark matter in the Universe, perhaps in the form of galactic haloes (Rogstad and Shostak, 1972; Ostriker and Peebles, 1973; Roberts and Rots, 1973; Ostriker et al., 1974; Ostriker and Thuan, 1975; Thuan et al., 1975; White and Sharp, 1977; Hegyi and Gerber, 1977; Miller, 1978; Karachentsev, 1978; Einasto, 1978; Gunn, 1978; Spinrad et al., 1973; Wesson, 1978g, 1979c), may render the larger value of ρ_U plausible.

The second term in (9.1) represents the acceleration of any particle due to the expansion of the cosmological substratum. As in the Weyl equation of motion (2.4), the term in $H\dot{r}$ may help to explain the apparent discrepancy between the masses of clusters of galaxies as deduced from the assumption that they obey the virial theorem as based on the usual Newtonian equation of motion ($\ddot{r} = -GM/r^2$) and the masses as deduced from observations (Ambartsumian, 1961; Rood, 1965, 1969, 1970, 1974a, b, c, d; Rood and Baum, 1967; Rood et al., 1970, 1972; Noerdlinger, 1970; Field and Saslaw, 1971; Rood and Sastry, 1971; Chincarini and Rood, 1974; Antonav and Chernin, 1977; Wesson and Lermann, 1977a; Wesson et al., 1977; Wesson, 1977b, 1978g; Carr, 1978; Rood and Dickel, 1978). Ginzburg (1975, 1976) has discussed this and related problems in astrophysics that appear to require the introduction of new physics. An analysis by Lewis (1976) has shown that in principle an equation of motion which has a term incorporating a secular relaxation effect, such as might arise from a slow decrease with time of the value of G or from a cosmological acceleration as in (2.4) and (9.1), can help to explain the missing mass (virial discrepancy) problem. Lewis has used a computer program with $G(t) = G_0(t_0/t)$ and $t_0 = 1.5 \times 10^{10}$ yr to study the evolution of galactic orbits with a time-varying G. He has shown that the virial discrepancy can be removed for groups of galaxies under this condition, and that after a group or cluster ceases to be gravitationally bound the relaxation effect introduces an extra expansion velocity characterized by $\dot{r}/r \simeq -\dot{G}/G \simeq 70$ km sec^{-1} Mpc^{-1} using Van Flandern's value of $\dot{G}/G$ (Van Flandern, 1975). The fact that this latter figure is approximately equal to H_0 ($\simeq 75$ km sec^{-1} Mpc^{-1} $\simeq 2.5 \times 10^{-18}$ sec^{-1}) is typical of all processes which, like those described by (2.4) and (9.1), introduce non-Newtonian dynamical effects that operate on cosmological time scales ($t_0 \approx H_0^{-1}$). Prokhovnik (1970a) has commented on other astrophysical consequences of the equation of motion (9.1), especially as they have to do with gravitational waves and a possible identification of the cosmological substratum of the model with the frame in which the 3 K microwave background appears to be fully isotropic.

As a theory of gravity, Prokhovnik's cosmology is not complete because it lacks field equations of the type which have been so successfully tested in the case of general relativity. (In this respect, Prokhovnik's theory and others like it to be examined below are similar to Milne's theory mentioned in the main text: they are really statements of symmetry principles with which any subsequently proposed field equations would have to be consistent.) Nevertheless, a theory of the Prokhovnik type is justified, as far as it goes, if (as in this case) it can be connected up with observable consequences that can help to direct attention to topics in gravitational theory which might profitably be investigated in more detail.

(4b) Segal (1972, 1976a) has given a cosmological theory of gravity, which he calls a chronometric theory and which is based on a variant of special relativity (Segal, 1974). It grew out of a study of the 15-parameter conformal group in a cosmological context (Segal, 1972); and while the modern form of the theory is essentially a cosmology (Segal, 1976a, b, c), the use of the conformal group has a long history both in particle physics and in theoretical studies of space-time structure (see below; the connection with particle physics and the origin of the term chronogeometry was reviewed by Segal (1972), along with the astrophysical implications of these subjects). The space-time of the cosmology is locally identical with Minkowski space-time, but is globally different. Astronomical objects in a Universe described by the chronometric theory have finite redshifts, and as far as astrophysics is concerned most of the content of the theory concerns how the redshift z depends on the distance r of the source from the observer. In this context the theory as it was originally presented was scarcely tenable: Segal (1972) tried to make a case for identifying the redshifts predicted by the conformal group with the redshifts of the galaxies, denying that the latter are majoritively Doppler shifts due to the expansion of the Universe. Arguments of this type are common in astronomy, and are rightly viewed with suspicion. A discussion of the main arguments for the existence of redshift phenomena that are incompatible with conventional big-bang cosmology is given in Chapter 11. One may quote here part of the last sentence of that chapter: There is no evidence of any kind at the present time that leads one to doubt that galactic redshifts are understandable as the result of the Doppler effect acting in an expanding, general-relativistic Universe.

This must make one automatically sceptical of Segal's theory and others like it in which the redshift is not due mainly to the Doppler effect. (A hypothesis of Bellert, (1969, 1970, 1977), which is similar to that of Segal in that it accounts for the redshift in terms of a particular choice of time coordinate in a static space-time, leads to a time-variability of the velocity of light c as that quantity is involved in astrophysical processes such as supernovae explosions, and so can hardly be acceptable in view of the contents of Chapter 11. Criticisms of Einstein's special relativity by Jánossy (1971), Brillouin (1971; see also Woodward and Crowley, 1973) and Dingle (1973) are not regarded as valid by most workers, and in any case the usual Lorentz transformation formulae, as opposed to formulae of a similar but more general type (Alway, 1969), were confirmed as correct by Strnad (1970). The Lorentz formulae themselves have been tested to great accuracy (see below), and Cialdea (1972) has confirmed the result of the classical Michelson/Morley experiment using a laser technique that can measure velocities to an accuracy of about 1 m sec^{-1}). Segal's later presentation of the chronometric theory (Segal, 1976a, b, c) was more refined than the earlier one, but even though some small redshifts might arise from peculiar motions of the sources, the major part of the effect is still supposed to be non-Doppler in origin. Physically, the redshift is supposed to be due to the fact that, whereas photons in the laboratory are localized, those emitted from distant galaxies are not localized (the difference is due to the existence of two time operators in the conformal group). While the theoretical basis of the theory does not seem to be amenable to test, the expression for the redshift can be predicted to be

$$z = \tan^2 (ct/2R_U) = \tan^2 (r/2R_U), \qquad (9.2)$$

where t is time, r is distance and R_U is a constant which may be described as the 'radius' of the Universe. This expression, for objects that are not at too great distances, reduces to $z \propto r^2$, and this can be tested.

In this regard, work on the redshift problem by Segal (1975) was aimed at providing a validation of the chronometric theory by studies of the z/r relation and the magnitude/redshift relation for galaxies. An analysis of low-redshift galaxies was carried out by Nicoll and Segal (1975), and a more comprehensive statistical study of how galaxy redshifts depend on distance yielded the result $z \propto r^p$ with $p = 2$ (± 0.2) for galaxies with $z \lesssim 0.02$ (Nicoll and Segal, 1978). This result has been bolstered by Segal (1978a) who has termed as erroneous a previous criticism by Fairchild (1977) of the way in which the z/r relation is presented in Segal's version of the chronometric theory. Segal (1978b) has also presented an analysis of medium-redshift sources from Markarian's list of Seyfert-like galaxies (mean $z \simeq 0.04$) which confirms the result $z \propto r^2$ found by Nicoll and Segal. This at first sight seems to be convincing support for the chronometric theory, which as noted predicts that $z \propto r^2$, and a death blow for all of conventional astronomy, which is based on Hubble's law (velocity of recession $\dot{r} = Hr$) and the connected relation $z \simeq \dot{r}/c \simeq Hr/c \propto r$. (A parameter which is analogous to H can enter into Segal's theory via R_U in (9.2) above, and Köhler (1978) has tried to explain what appear to be super-luminal velocities in sources such as QSOs by a suitable choice of R_U; but he has found that real velocities greater than c can only be avoided for a value of $R_U \simeq 50$ Mpc, which is unreasonably small.) However, the claim that $z \propto r^2$ has been around for some time (Hawkins, 1960, 1962a, b), and does not necessarily carry the fatal conviction attributed to it by the Segal theory. Indeed, the effect is probably real, but is restricted to low-z sources and can largely be explained as a dynamical effect of the Local Supercluster on the motions of nearby galaxies (De Vaucouleurs (1972a); the z/r relation is further discussed by Wesson (1979a, b) in relation to the magnitude/redshift relation for galaxies in an inhomogeneous Universe model). Part of it could also be due to luminosity selection effects that might be present in samples of bright, nearby galaxies (Brecher and Grunsfeld, 1978). Certainly, the conventional relation $z \propto r$ and Hubble's law appear to be valid when one considers galaxies in remote clusters, the apparent magnitudes of which give a Hubble (m/z) diagram with the conventional slope (5) out to $z \simeq 0.5, m \simeq 19$ (Sandage et al., 1976; Peebles, 1978a). Even Nicoll and Segal (1978) have admitted that the conventional relation $z \propto r$ is compatible with galaxy redshift data for high-redshift ($z \simeq 0.1$) sources.

Clearly, the status of the chronometric theory as regards its comparison with observation is shaky to say the least. On the other hand, it must be admitted that the use of the conformal group as a basis for a theory of gravitation is appealing. There are several such theories besides Segal's and these are examined in Chapter 10 along with other aspects of group theory as it relates to gravitation.

(4c) Ni has developed a metric theory, based on a Lagrangian, which possesses a preferred frame with conformally flat space slices (Ni, 1973). With reasonable choices of the functions involved and of a cosmological model, the Ni theory has the same post-Newtonian limit as general relativity, and can therefore be made to agree with all relevant tests which have been made to date. (It is necessary in general that a theory of gravity both be a metric theory and have a correct Newtonian limit if it is to be successful, since Will (1974) has shown that the gravitational redshift effect can only be explained in terms which are independent of the nature of the atomic clocks concerned by theories which are metric-based. This conclusion is widely regarded as rendering all non-metric theories of gravity inadmissible. But it should be mentioned on the other side that Schiff (1960) has argued that the gravitational redshift and light deflection effects can be derived from non-metric considerations, at least to first order; while Roxburgh and Tavakol (1975) have shown that the question of whether a theory is expressible in metric format depends among other things on finding a suitable space, which may not necessarily be Riemannian, in which it can be geometrized. Ni's theory, of course, avoids these ambiguities since it uses a conventional Riemannian space-time.) The Ni theory has been discussed by McCrea (1973) and is interesting in that, unlike most preferred-frame theories, it is guaranteed by construction to have post-Newtonian parameters satisfying the conditions for acceptability derived by Will (1971). It is also in agreement with conditions on conservation laws for energy, momentum and angular momentum in metric theories discussed by Lee et al. (1974). As noted by Ni (1973), further development of the theory will depend on how cosmological models deduced from it agree with observation, and on the properties (polarization, intensity and speed of propagation) of gravitational waves in models based on the theory.

(4d) Rastall's theory (Rastall, 1977a, b, 1978) developed out of a super-Newtonian theory in which there was assumed to be a preferred coordinate system of the type familiar from Newtonian mechanics and special relativity (Rastall, 1968, 1975). In its original form the theory employed a metric that was taken to be a function solely of the gravitational potential, but this condition was later relaxed (Rastall, 1976), allowing the gravitational parameter G to become time-variable. Subsequently (Rastall, 1977a, b), the theory was modified by deriving new field equations from a Lagrangian constructed in a Minkowski frame. In its later form (Rastall, 1976, 1977a, b, 1978), the theory is described by a metric that depends on a scalar function and a vector field. The theory in this form can be tested astrophysically by calculating the upper limit for the mass of a collapsed object such as a neutron star and comparing this to the masses of observed objects. The theoretical upper limit on the basis of Rastall's theory is 4.5 $M_\odot$ if the speed of sound in the star is nowhere greater than the speed of light (Rastall, 1977c). This is somewhat larger than figures given by similar calculations using Einstein's theory; and since some neutron stars appear to have masses close to or perhaps exceeding the values typical of general relativity, Rastall's theory might hope to gain support from more exact observations of compact astrophysical objects. (This topic will come up again during a discussion of Rosen's bimetric theory: see (5a) below.) On a larger scale, cosmological solutions of the theory which are homogeneous and isotropic can be derived

(Rastall, 1978). There are no solutions corresponding to closed Universes, while the open-Universe solutions give a value of the deceleration parameter in the range $-0.03 > q_0 > -0.45$. While the theory may be astrophysically viable, and appears to have an acceptable parameterized post-Newtonian limit (Rastall, 1977a, b; Smalley, 1978), Rastall (1978) has himself remarked that further development of a theory of this type should logically await the availability of more data on the observable parameters that are involved.

(4e) Clube has discussed the nature of gravity within the framework of a theory in which there is assumed to be a preferred Euclidean frame for space-time (Clube, 1977). This assumption involves an arbitrary redefinition of particle masses and the velocity of light c, and is apparently based on a discussion by Atkinson of light deflection, particle motion and the redshift of spectral lines near a large stationary mass situated in a flat metric background (Atkinson, 1963; see also Atkinson, 1965a, b). The theory seems *ad hoc*, and it is not clear that it has any limit in which its predictions agree with the observationally verified ones of general relativity. Clube (1978) has discussed some astrophysical data that conventional cosmology has difficulty in explaining and which might be understandable on the basis of his new theory. But one such effect – a possible expansion of the Galaxy at about 40 km sec^{-1} at the Sun's distance from the centre – has been shown not to exist by Ovenden and Byl (1976; see also Chapter 6). There seems, therefore, to be as little observational as there is theoretical justification for Clube's theory.

Another class (5) of theories comprises those which have what may loosely be termed a conventional space-time structure but which are inequivalent to general relativity.

(5a) Freund *et al.* (1969) have formulated a finite-range gravitational theory. It is a tensor theory, and can be formally derived from a Lagrangian that is based on considerations concerning the energy-momentum tensor for the matter in the Universe. But because the existence of a finite range of action for gravity leads to the abandonment of the principle of general covariance (on which general relativity is founded), the theory cannot be given an unambiguous geometrical interpretation. In spite of this, the assumption that there exists a model Universe with a metric which is analogous to that of the Robertson/Walker model in Einstein's theory, and in which the matter is taken to be a fluid with zero pressure, enables a cosmologically interesting solution of the field equations of the new theory to be found. This solution, although it can be said in a loose way to have a space-time structure similar to that of the Friedmann models at least locally, has some distinctly non-Friedmannian properties on the large scale. The force of gravity now has a finite range of the order of the Hubble radius of the Universe. Dynamical properties of the model therefore depart from those one would expect from general relativity on the cosmological scale, although locally the differences between the finite-range theory and Einstein's theory are extremely small. The model is an oscillating one, with a cosmological redshift that depends on the analogue of the scale factor of conventional theory. It would seem that agreement between theory and observation can be attained as far as the redshift is concerned, if not by the particular model considered by Freund *et al.* (1969) then by another one based on their theory. The latter, while it may be astrophysically acceptable, has perhaps its strongest claim to consideration in that the graviton, which is responsible for the finite range of gravitation in the theory, has of necessity a finite mass (as opposed to the situation in general relativity, where gravitation has an infinite range and the graviton has zero mass). By the Uncertainty Principle, an upper limit of $\hbar H/c^2$ can be set for the mass of the graviton, which is only of the order of 10^{-66} gm and so very small. The existence of a finite if small graviton mass makes the finite-range theory readily quantizable, which is a desirable property from the viewpoint of particle physics.

The possibility that gravity has a finite range has been discussed in astrophysics mostly in connection with the clustering of galaxies. Zwicky inferred a possible weakening of gravity at large distances from his belief in the non-existence of superclusters (Zwicky, 1961; see also Gerasim, 1969 and Chapter 10). This belief was erroneous, as progress in extragalactic astronomy has shown that superclusters unquestionably exist (see, e.g., De Vaucouleurs, 1971; Groth *et al.*, 1977; Chapter 11). There is therefore no observational justification from data on scales of order 10 Mpc for assuming that gravity might have a finite range.

On other scales, Finzi raised the question of whether Newton's law is valid over all cosmic distances, and suggested that a force law with a fall-off that is slower than the Newtonian one might help to resolve some astrophysical problems and especially that of the virial discrepancy in clusters of galaxies (Finzi, 1963a, b). This suggestion was soundly criticized by Yabushita (1964), who pointed out that it would lead to unacceptably large dispersion velocities for galaxies in clusters and result in other dynamically unacceptable consequences. Further, Rood (1974c) showed that a slower-than-Newtonian fall-off in the law of gravitational attraction would not in practice provide an explanation of data concerning the missing mass problem in clusters. These results show that there is no justification for the idea of a finite range for gravity as far as that hypothesis affects astrophysical systems on scales of order 1 Mpc.

As far as smaller scales are concerned, various workers have deduced that the Newtonian gravitational parameter G may depend to a small extent on the value of the gravitational potential (see Wesson (1978a) for a review). Finzi (1962) has discussed ways in which such a space-variability of G could be detected inside the Galaxy by observations of white dwarfs. A possible dependence of G on the potential could be tested for on the scale of the Solar System by Doppler tracking of natural and artificial satellites (Finzi, 1968). However, no data yet exist which indicate that gravity is not an inverse-square law or that G is space-variable on Galactic and planetary scales.

On the contrary, some data exist which show that G is not variable in space at least on short scales. At laboratory distances, a departure from a perfect inverse-square law was reported by Long (1976), but the results of his experiments were not confirmed by the subsequent work of Newman (1977; see also MacCallum, 1977). A possible scale-dependence of the force of gravitation of the type proposed by Long was also effectively ruled out by Blinnikov (1978). He noted that departures from conventional theory would cause modifications of the mass/radius relation for degenerate stars like white dwarfs. The agreement between observations of such objects and calculations based on conventional theory allowed Blinnikov to conclude that the laboratory-determined value of G and its value on the scale of stellar bodies agree to within 10 percent.

This last negative result completes a list of similar results which, in sum, tend strongly to uphold the validity of the conventional gravitational law over the range of distances 10–10^{25} cm for which data are available with which to test it.

The conclusion that there is no evidence to support a possible finite range for gravitation does not necessarily invalidate the theory of Freund et al. (1969) because, as noted, their presentation of that theory concentrated on a form in which departures from conventional theory only become dominant on cosmological scales of the order of $cH^{-1} \approx 10^{28}$ cm. But it must nevertheless be stated that the validity of conventional theory on smaller scales does give collateral support to the accepted view that gravity is an infinite-range force, while the lack of any decisive way in which the form of the law of gravitation can be tested on scales of order 10^{28} cm is a conceptual drawback to the finite-range theory. Thus, although that theory may be formally acceptable, it lacks observational justification.

(5b) Rosen, believing that a homogeneous, isotropic model of the Universe possesses a preferred frame of reference and a preferred time coordinate, analogous to the absolute space and time of Newton (Rosen, 1969a, b), abandoned the concept of covariance on which general relativity is based and formulated a non-covariant theory of gravitation (Rosen, 1971a, b). It is a bimetric theory, meaning that massive bodies produce curvature fluctuations which are superimposed on a constant, flat metric background. The metric coefficients can thus be written as

$$g_{ij} = \gamma_{ij} + h_{ij}, \tag{9.3}$$

where $\gamma_{ij} = \eta_{ij}$ is the flat-space tensor (up to a coordinate transformation) and h_{ij} describes the matter present. The metric tensor h_{ij} is Riemannian, and the condition $\gamma_{ij} = \eta_{ij}$ makes it particularly easy to formulate the equations of the theory and work out its observable consequences (Rosen, 1973). The background does not affect the equation of motion of a test particle or the trajectory of a light wave near a massive body like the Sun, so the three classical tests of general relativity have the same status in Rosen's theory as in Einstein's. However, in Rosen's theory, gravitational waves can be emitted by spherically-symmetric matter configurations (Rosen, 1971a; Stoeger, 1978), in contradistinction to the situation in general relativity.

It should be noted that the differences between Rosen's theory and Einstein's theory are due to the field equations involved (see below) and not directly to the break with the covariance principle or to the adoption of a bimetric nature for space-time. The attaching of special significance to cosmic time had earlier led Szekeres to propose a reformulation of general relativity that has several features in common with Rosen's theory (Szekeres, 1955; Kantor and Szekeres, 1956), and Katz has adopted a similar standpoint (Katz, 1974; see below). The bimetric condition is also consistent with a range of metric theories including general relativity (see Nelson, 1972), and Rosen (1940a, b, 1963) had previously studied various aspects of the flat-space metric in Einstein's theory. The significance of the covariance and bimetric conditions lies in the fact that a break with the one and an adoption of the other represented the original motivation that led to the proposal of the new theory.

In its original formulation, the bimetric theory was in agreement with the usual relativistic tests even though it did not incorporate covariance (Rosen, 1973). However, Rosen later changed his stance somewhat on the question of covariance and decided to incorporate a formal agreement with the covariance principle into the new theory (Rosen, 1974). This he did by allowing γ_{ij} in (9.3) to be unrestricted, although in practice the condition $\gamma_{ij} = \eta_{ij}$ was sometimes retained in later developments of the theory for the sake of convenience. Besides being in accordance with the principle of covariance, the bimetric theory was also shown to be in explicit agreement with the Principle of Equivalence by Goldman (1976). He derived expressions for the inertial and gravitational masses of a static, spherically-symmetric body in the theory and showed that they were equal, irrespective of the equation of state adopted. The field equations of the covariant bimetric theory were derived from a Lagrangian composed of parts describing the background field (γ_{ij}), the field describing the usual gravitational interactions of matter (h_{ij}) and a field describing other, non-gravitational forces (Rosen, 1974). The problem of a massive body and its static, spherically-symmetric field was examined, along with the weak-field case of the field equations. (It was claimed by Yilmaz (1975) that the equations of Rosen (1973) were inconsistent, but it has been pointed out (Rosen, 1976) that this conclusion was only due to a mistaken use of certain auxiliary conditions which were not employed in the later (Rosen, 1974) version of the theory.) The newer covariant form of the bimetric theory, like its older non-covariant form, is in agreement with the classical relativity tests, and it is only in the realm of astrophysics and cosmology that one can expect to distinguish observationally between the bimetric theory and general relativity.

One way of testing the Rosen theory astrophysically is by calculating the upper mass of a neutron star, as in Rastall's theory outlined previously. This was done by Rosen and Rosen (1975) who found a figure of 8.1 $M_\odot$ using an approximate equation of state. General relativity, with the same equation of state, gave an upper limit of only 1.5 $M_\odot$, which is typical for reasonable equations of state in Einstein's theory (Hartle and Thorne, 1968). For comparison, scalar-tensor theories like the Brans/Dicke theory give upper masses similar to those predicted by general relativity (Salmona, 1967; Matsuda, 1972; Yokoi, 1972). The high value of 8.1 $M_\odot$ is interesting in view of indications, noted in (4d) above, that there may exist objects with masses exceeding the Einstein upper limit (Rastall, 1977c). The binding energy of a cold neutron star in the bimetric theory was also found to be considerably larger than that given by general relativity in several models of static spheres of matter composed of baryons studied by Goldman and Rosen (1978). This difference can manifest itself as a difference in the predicted radii of neutron stars in the two theories, and together with the mass difference this means that neutron stars provide two ways of differentiating between Rosen's theory and Einstein's theory.

On the cosmological scale, a flat ($k = 0$) homogeneous, isotropic cosmological solution has been found in the Rosen theory by Babala (1975). The solution behaves like the Lemaître ($\Lambda \neq 0$) model of general relativity, a universal repulsive force being present directly as a consequence of the Rosen field equations. (This supports a conjecture by Wesson (1978a) that the background metric in the Rosen theory plays the same role as the Λ-term in

Einstein's theory.) However, the early stages of a model Universe in the Rosen theory can be similar to the early stages of an expanding model in general relativity: a model Universe filled with isotropic radiation in the bimetric theory begins its expansion from a singular state with infinite energy density (Goldman and Rosen, 1976), so there should be no problem in accounting for the origin of the 3 K black-body radiation field. A negatively curved ($k = -1$) cosmological model which is homogeneous and isotropic has also been studied in the bimetric theory (Goldman and Rosen, 1976, 1977). This model is characterized by a predicted change of the gravitational parameter at the (observationally acceptable) rate of $\dot{G}/G \simeq + 4 \times 10^{-12}$ yr^{-1}, and contains matter with negligible pressure. (The Rosen theory was not primarily formulated as a way of obtaining G-variability, but as discussed by Lee et al. (1976) such variability is consistent with a cosmological application of the theory.) For astrophysical situations in which the pressure is not negligible in comparison to the energy density of the matter, the Rosen theory appears to predict an asymptotic incompressible state for collapsed systems (Rosen and Rosen, 1977). This is because there is no limit in the Rosen theory for the mass of a compressed body, so the final stage of gravitational collapse need not be a black hole as it is in general relativity. Rosen (1977a) has discussed the status of the bimetric theory in relation to observational data from cosmology, reaffirming his earlier statement (Rosen, 1973, 1974) that prsently available data are compatible with the bimetric theory as well as with general relativity. Rosen (1977a) was mainly concerned with replying to work by Lee et al. (1976) on the post-Newtonian limit of the bimetric theory. The latter showed that, with one exception, the post-Newtonian parameters that describe Rosen's theory are identical to those of Einstein's theory. The exceptional parameter can be fixed by cosmological considerations, and relates the value of G to the structure of the Universe. Lee et al. (1976) showed that for a plausible cosmological model, G has its observed value and the outstanding post-Newtonian parameter is zero, making the post-Newtonian limit of the bimetric theory identical to that of general relativity. Rosen (1977a) has extricated himself from this embarrassing situation by considering cosmological counter-examples in which the parameter involved is non-zero so that the bimetric theory retains its individuality.

It has been suggested (Will and Eardley, 1977) that Rosen's theory is in conflict with observation in that it predicts an excessive rate of emission of dipole gravitational radiation from the binary system PSR 1913 + 16. (This is in contrast to general relativity, where gravitational radiation is quadrupole in nature.) A loss of energy by gravitational radiation at a large rate would cause excessive changes in the orbital period of the binary system, contrary to an observed upper limit on period changes of $|\dot{P}/P| \lesssim 1.2 \times 10^{-7}$ yr^{-1}. Rosen (1978) has responded to this attack by pointing out that there are some basic arguments (Rosen, 1977b) for believing that in the bimetric theory, as in general relativity, a system cannot lose or gain energy by the emission of gravitational radiation.

One sees that while the status of Rosen's theory continues to be debated, it cannot yet be ruled out either on the basis of classical Solar System relativity tests or astrophysical observations.

(5c) Omote (1971, 1974) has formulated a gravitational theory which is locally scale invariant, meaning that all predictions of the theory are invariant under transformations of the form $x^i \to \rho(x)x^i$ for the coordinates, where ρ is an arbitrary function. It can be mentioned here that the same condition has also been used by Freund (1974), whose locally scale-invariant theory appears to be essentially the same as that of Omote.

A discussion of scale invariance and conformal invariance and their use in physics has been given by Gross and Wess (1970). They note that, although the two concepts are logically distinct and have their own descriptions in terms of group theory, it is often the case that scale invariance implies conformal invariance. In gravitational theory, scale invariance is a concept usually applied to coordinates while conformal invariance is a concept usually applied to fields or the metric. When applied to intervals, a conformal transformation implies that the ratio of two lengths, possibly in different directions but at the same point, does not change under the transformation (Hoyle and Narlikar, 1974). Conformal transformations, which are often regarded as a kind of generalization of scale transformations, correspond physically to transitions to accelerated coordinate frames (Hill, 1945a, 1947). The kind of invariance employed by Omote has its natural description in terms of the type of metric background considered by Weyl (1919, 1922; see also Chapter 2; scale invariance is discussed further in Chapter 8 and conformal invariance is discussed further in Chapters 2 and 10). Omote, realizing that scale transformations of the noted type are equivalent to gauge transformations of the metric $g_{ij} \to \beta^2(x)g_{ij}$, has formulated a locally scale- (or gauge-) invariant theory using the Weyl geometry.

He has done this by constructing a (rather complex) Lagrangian that is invariant under all coordinate and scale transformations and which contains a scalar field ϕ which is similar to that of the Brans/Dicke theory. Gravitation depends on the scalar field $\phi(x)$ as well as on the conventional metric tensor $g_{ij}(x)$. In addition, the gauge invariance brings in a gauge field which obeys an equation similar to Maxwell's equation for the electromagnetic tensor.

(Indeed, the corresponding field in Weyl's original geometry was identified by Weyl with the electromagnetic field, but this led to conflicts with observation that were the cause of a lapse in interest in field theories of the Weyl type. It is only in recent years that the Weyl geometry has experienced a revival.) In the Omote theory, the role of the gauge field is not completely defined, except that it can interact with neutral fields as well as charged fields and so is not to be identified with electromagnetism. The three fields (tensor, scalar and gauge) give the theory great flexibility. In general form it is approximately equivalent to the Dirac (1973) theory, while the variability of the scalar field (ϕ) brings in a variation of the Newtonian gravitational parameter G as in the Brans/Dicke theory. In the latter regard, though, the Omote theory is better formulated because, unlike the Brans/Dicke formalism, the scale-invariant theory is obliged to have a variable ϕ since a constant ϕ would break the scale invariance. (A constant ϕ also corresponds to general relativity.) The variation of ϕ (with, for example, cosmic time) ensures that no dimensional constants appear in the theory (compare this with the discussion of the work of Deser (1970) in the first part of Chapter 8). The variability of the scalar and gauge fields can be adjusted as desired, so the Omote theory can be brought into acceptable agreement with observation by a suitable choice of parameters.

(5d) Katz has presented a theory of gravity based on the existence of a preferred cosmic time (Katz, 1974; compare (5b) above). It is a metric theory, which admits cosmological solutions that bounce without going through a singularity at the expense of introducing a new coupling parameter. Under some conditions, the field equations reduce to those of general relativity, and overall the Katz theory appears to be in agreement with presently available observations.

The Katz theory is not the only account of its type. Rosen (1969c) also realized that homogeneous, isotropic models of the Universe could oscillate without going through a singular state provided one introduces a scalar cosmic field that has the characteristics of a negative pressure. (Spin and torsion have qualitatively similar effects, and Trautman (1973) has shown how gravitational singularities may be averted in space-times with rotation.) A related procedure had earlier been used by Pachner (1965), who introduced a stress that depended on the curvature of space-time, thereby avoiding the singularity in an oscillating isotropic Universe model and at the same time including continuous creation of matter within the confines of general relativity. Rosen (1969c) pointed out that if one adds a term corresponding to a particular type of negative pressure to Einstein's equations, one obtains the C-field theory of Hoyle and Narlikar given that there is conservation of matter. The Hoyle/Narlikar C-field, which is discussed in Wesson (1978a), was a forerunner of the theory discussed in Chapter 2. It was a cosmological theory involving continuous creation of matter, and while it was later superseded as a cosmology it nevertheless allowed some interesting results to be obtained that have relevance to other continuous-creation cosmologies where matter is produced at the same rate everywhere. Using a model of continuous creation due to a scalar field, Hoyle and Narlikar (1963) showed that creation tends to smooth out density irregularities in the matter of the cosmological medium, so helping to account for the rough isotropy of the Universe. Continuous creation also has the effect of reducing any cosmological rotation, so helping to explain why the local inertial frame appears to be the one with respect to which distant matter in the Universe is non-rotating.

However, McCrea (1951) had introduced continuous creation of matter in accordance with Einstein's theory some time before the related work of Rosen, Pachner and Hoyle/Narlikar referred to above. McCrea showed in effect that continuous creation can be accounted for with the equations of conventional general relativity plus an equation of state that includes a negative pressure ($p < 0$) and a positive matter density ($\rho > 0$). In the case where the equation of state is $p + \rho c^2 = 0$, the metric of a general relativistic model with continuous creation is just the de Sitter metric, and in general one can find solutions of Robertson/Walker type in which the created matter can be viewed as the work equivalent of the stress term ($p < 0$) acting in an expanding Universe.

In McCrea's model, continuous creation is introduced in a quite natural way. The $p < 0$ condition would in particular appear to be a more natural way of accounting for creation than the postulate of a special creation or C-field (Hoyle, 1949a, b). The latter postulate was used in the original steady-state theory, which, despite its shortcomings, is still often regarded as the archetype of all Universe models that incorporate continuous creation. (A recent attempt to resurrect the old form of the steady-state Universe by Surdin (1978) introduces an all-pervasive electromagnetic field at the absolute zero of temperature, stochastic fluctuations of which account for the continuous creation of matter and the existence of the microwave background radiation. Zero-point fluctuations of an electromagnetic field of the type involved here may, as suggested by Hobart (1976), be connected with the degree to which the Universe is a perfect absorber in the sense of Wheeler/Feynman electrodynamics, and so with the value of Planck's constant.) It should be noted that in continuous creation processes that are consistent with metric theories of gravity, the new matter does not come *from* anywhere: it just appears. It is not, for example, imported from another Universe since (at least as far as general relativity is concerned), while this is theoretically possible (Wheeler, 1962; Carmeli et al., 1970) in practice it is impossible because of instabilities in the metric of the space-time bridge (Birrel and Davies, 1978). The matter which is created in the McCrea model appears uniformly everywhere in space at a rate of order 10^{-44} gm cm^{-3} sec^{-1} if p is identified with a cosmological stress field. But in principle matter in models of this type could also be created preferentially in places where matter is already most dense (as in the Dirac ($\times$)-model; Bondi (1961) considered it unlikely that stars could be doubling their masses on a stellar time scale of 3×10^9 yr due to continuous creation at a rate of order 10^{-17} gm/gm sec, as might be expected for creation where matter is already most dense). In either case, such continuous creation would be expected to have serious effects on the evolution of galaxies and clusters of galaxies (McCrea, 1950). No evidence exists which can rule out uniform creation everywhere at the rate noted above; but for creation at a rate proportional to mass an upper limit exists of order 10^{-23} gm/gm sec, which, considered in conjunction with other possible consequences of secular mass changes in astronomical bodies (Wesson, 1973, 1977a, 1978b) places constraints on the admissibility of continuous creation of this type.

Restrictions on the admissibility of continuous creation such as the upper limit just quoted might not apply to situations like that envisaged by Shäfer and Dehnen (1977a, b), where there is simultaneous creation of particles and antiparticles out of the vacuum, due in this case to fluctuations of the quantized Dirac field embedded in an expanding Robertson/Walker background. The model considered by Shäfer and Dehnen (1977a, b) for $k = +1$ and $k = 0, -1$ cosmologies respectively, accounts for the origin of the matter in the Universe essentially by introducing a continuous creation term due to particle physics. This creation term is proportional to H^2 ($H = \dot{S}/S$, where the scale factor S depends only on the time), and acts in opposition to the usual depletion term ($\dot{\rho}/\rho = -3\dot{S}/S$) due to the expansion of the Universe. While the model they consider gives an m/z relation in approximate agreement with the observed one, it is problematical in that it presumably implies that half the energy of the Universe is residing (somewhere) in the form of antimatter.

The status of baryon-symmetric big-bang cosmology (in which there are equal amounts of matter and antimatter) has been discussed by Steigman (1974) and Stecker (1978). Although such a cosmology cannot be ruled out on observational grounds, one gets the feeling that it is inherently implausible. (After all, it takes a lot of energy to create antimatter on the Earth.) On theoretical grounds, baryon-symmetric cosmology may be untenable if the early Universe was dominated by a unified gauge theory of particles interactions. Yoshimura (1978) has shown that, according to at least one such theory which incorporates CP and T violation, the antiprotons which might have been present in the big-bang fireball would have tended to be eliminated with time. This model accounts for the eventual dominance of matter protons over antimatter protons, as apparently observed at the present epoch. It also predicts a photon/baryon ratio of order 10^9, which compares well with the observed value of order 10^8. The production of a net baryon density by processes that violate baryon-number conservation has also been considered by Weinberg (1979). He has pointed out that the model treated by Yoshimura cannot produce an appreciable baryon density if all relevant channels are taken into account, but that departures from thermal equilibrium in an expanding fireball can in general account for a significant baryon density and lead to a photon/baryon ratio of order 10^9. The principle of Yoshimura's argument, if not its details, would therefore seem to be valid. The implication is that the apparent excess of baryons over antibaryons in the observed Universe can be understood as a natural consequence of particle processes in the big bang. (Other ways of accounting for the photon/baryon ratio have been mentioned in Chapter 3 previously, while bouncing Universe models (Gold, 1962; Davies, 1972, 1974; Wesson, 1978a) might account for it as accumulated, thermalized starlight.) There are seen to be reasons for being sceptical about the presence of large amounts of antimatter in the Universe that are of big-bang vintage; while if antimatter is being created at the present epoch, the process must be occurring in a way which ensures that the antimatter is kept isolated from the matter of the Galactic neighbourhood. It seems more likely that if continuous creation occurs, then it is in the form of purely normal (not anti-) matter, and it has been seen that a negative pressure term can account for this quite satisfactorily.

The fact that a negative pressure in general relativity can achieve what other theories do in a more complex manner (compare Dirac's theory (2a) discussed in Chapter 2 and elsewhere), should act as a caution against a too-ready desire to abandon Einstein's theory in favour of one of its more modern rivals. McCrea (1978) has pointed out that recently-proposed theories in which G is time-variable might logically be better formulated as accounts in which masses vary with time. This certainly holds in general relativity, where masses can be made time-dependent by introducing a pressure (see above and Wesson, 1978a). It also holds in a way for the simpler case of Newtonian theory. The logicality of G-variability thus has relevance to attempts such as that of McVittie (1978) to construct Newtonian G-

variable models. There is a formal correspondence between relativistic and Newtonian cosmology (Layzer, 1954; Evans, 1978); so, as far as it is sensible to discuss G-variability in the one case, it is also sensible to discuss it in the other case. McVittie's theory is one in which the Newtonian laws of motion retain their usual forms, with constant particle masses and $G = G(t)$. There exist solutions in which the matter distribution is homogeneous and in which $G \propto t^{-1}$. Bouncing models, of the type discussed above in connection with the work of Katz, Rosen and Pachner, in the McVittie (1978) theory have a structure in which each cycle lasts longer than its predecessor, until eventually G becomes so small that the matter becomes unbound and expands to infinity. This property of the Newtonian oscillating model avoids somewhat the philosophical objections which can be raised (Jaki, 1977) against bouncing models of the Universe in general.

(5e) Malin has proposed a group theory of gravity based on a use of the de Sitter group as a description of space-time and the Poincaré group as a description of the particles in it (Malin, 1974). The theory has been discussed by Wesson (1978a), and cannot be said to very well founded logically because it makes use of field equations which do not incorporate the principle of conservation of energy. Particle masses are time-dependent on a Hubble time scale in this theory, but as has been seen above in (5d) it is not really necessary to abandon Einstein's field equations in order to incorporate such a variation of masses into gravitational theory. Neither is there any positive evidence in favour of a time-variation of masses. (A secular extinction of mass in the Earth was discussed by Kapp (1960), while Karlsson (1971) has suggested on the basis of redshift data that the electron mass may be different from the conventional value, and perhaps quantized, in QSOs; but there is no reliable evidence in favour of either conjecture.) Indeed, one recalls from Chapter 4 that there is good evidence from astrophysics that the electron/proton mass ratio at least is not variable with time; and since the dynamical effects of mass loss and G-variability are equivalent, one infers from the contents of Chapter 6 that if masses are varying then they are doing so on a time scale $\gtrsim 10^{10}$ yr on the basis of Solar System experiments, and possibly on a time scale $\gtrsim 10^{12}$ yr on the basis of cosmological considerations.

However, while there may be little justification either theoretically or observationally for considering theories of gravity of the Malin type in which conservation of matter does not hold, it must be admitted that a variation of particle masses in general cannot be ruled out by experiment. Bekenstein (1977) has investigated the possibility that particle masses may be variable by constructing a relativistic dynamics for variable-mass particles moving in a space-time background for which Einstein's field equations are valid for a certain set of units (in which c, G and $\hbar$ are constants but in which rest masses vary). One case of this combination of particle dynamics and gravitation corresponds to a form of the Brans/Dicke scalar-tensor theory (Chapter 8), although Bekenstein's formulation of the equations involved is also of interest for other theories of gravity which do not obey the Principle of Equivalence in its strong form. Bekenstein has concluded that the classical Solar System tests of general relativity do not by themselves rule out the possibility of rest-mass variability, but that cosmological considerations may do so. Thus, one can say on the whole that variable-mass gravitational theory lacks justification and observational support, but cannot be definitely ruled out.

(5f) Goded has taken the condition $R_{ij} = 0$ (where R_{ij} is the Ricci tensor) as a basic condition to be obeyed by a free gravitational field. He has combined this condition with the usual conservation law $T^{ij}_{;j} = 0$ to obtain a theory of gravity that has something in common with both Newton's theory and Einstein's theory (Goded, 1975a, b; it has been pointed out by McVittie (1965) that the condition $R_{ij} = 0$ by itself, unlike the condition $R_{ijkl} = 0$ for the Riemann/Christoffel tensor, is not enough to ensure the flatness of space-time, so Goded's theory is not necessarily a Euclidean one). A suitable choice of parameters ensures that one recovers Newton's inverse-square law of gravitational attraction in the weak field limit, and the theory appears to be compatible with Solar System observations at some level (Goded, 1975a). In the cosmological realm, the use of the Robertson/Walker line element with a suitable choice of parameters yields motions of expansion in agreement with Hubble's law (Goded, 1975b), but the theory's implications for large-scale astrophysics are otherwise unknown.

9.3. An Intercomparison of Gravitational Theories

Having completed a review of modern theories of gravitation which might, at some level, be considered viable at the present time, it is instructive to compare them one with another. To sum up, there are five main classes of theories beyond the reference theory (0) of general relativity: (1) scalar-tensor theories, which basically means the Bergmann/Wagoner theory with its special cases of (1a) Jordan/Brans/Dicke, (1b) Nordtvedt and (1c) Bicknell/Klotz; (2) theories based explicitly on G-variability, meaning (2a) Dirac, (2b) Hoyle/Narlikar and (2c) Canuto *et al.*; (3) the vector-tensor theory of Hellings/Nordtvedt; (4) theories in which special use is made of a Minkowski frame, meaning (4a) Prokhovnik, (4b) Segal, (4c) Ni, (4d) Rastall and (4e) Clube; (5) other metric theories, including (5a) Freund *et al.*, (5b)

Rosen, (5c) Omote, (5d) Katz, (5e) Malin and (5f) Goded. This division into classes is, of course, somewhat arbitrary since several of the noted theories have properties in common despite the fact that they have been put into different classes depending on what aspects have been emphasized in their formulation. (For example, the Jordan/Brans/Dicke and Hoyle/Narlikar theories have similar mathematical structures, as have the theories of Dirac, Omote and Canuto *et al.*) Neither is there any guarantee that any of these alternative theories will prove to be a worthy successor to general relativity. On the contrary, the large number of alternatives that are available to Einstein's theory makes one sceptical of being able to isolate one special case as being favoured over its rivals.

Nevertheless, some comments can be made: (i) The discussion of this chapter has shown that theories of class (1) are only observationally acceptable in forms in which their consequences differ but slightly from those of general relativity. (ii) Of theories in class 2, the ($\times$)-model of the Dirac theory (2a) is in surprisingly good agreement with tests which have been made of it. The same cannot be said of (2b), although it is still viable (Wesson, 1978a). Theory (2c) is in agreement with all the tests which have been made of it so far, but its astrophysical status needs to be examined further. (iii) Theory (3) is mathematically acceptable (Ni, 1972) but lacks a compelling theoretical basis that could lead to its adoption in place of Einstein's theory. (iv) Prokhovnik's theory (4a) is attractive but incomplete. Segal's theory (4b) is not in accordance with the usually-accepted redshift law of astronomical objects, although the group on which it is based may well be of fundamental importance in future metric theories of physics. Theory (4c) is acceptable and is logically more attractive than the rather patchwork (4d) theory and the *ad hoc* (4e) theory. (v) The finite-range gravitational theory (5a) lacks observational justification, while the bimetric theory (5b) is a stronger contender for an acceptable theory of gravity, although one could wish on philosophical grounds that the latter was more different from (0) than it is. Theory (5c) is superior to (5d) and (5f) in that the first has a stronger foundation in scale invariance and is more general than the other two. Theory (5c) is also superior to (5e), since the latter lacks both theoretical and observational justification.

The comments of the preceding paragraph form the basis for the remarks made in Chapter 2 about the status of new theories of gravity and of the recommendations made in Chapter 7 for further work on non-Einsteinian theories of gravity.

CHAPTER 10

GROUP THEORY AND GRAVITY

10.1. Introduction

Although group theory has been of great success as applied to elementary particle physics, it has not been widely applied to gravitation. One notable exception concerns the conformal group on which Segal's theory of Chapter 9 is based. This group has a substantial history of usage in gravitational field theory, and in this chapter discussions are given first of the applications of the conformal group and related topics and then of more general ways in which group theory and gravity might be connected. Particular attention is paid to theories which, while unproven for the most part, are of interest because they attempt to unite elementary particle theory and electromagnetism with gravitation, in contrast to the theories examined in Chapter 9 which are primarily accounts of the gravitational interaction alone.

10.2. The Conformal Group and Related Topics

The elegant way in which the 15-parameter conformal group can provide a background for a theory of physical interactions had been realized long before Segal made use of it in his chronometric cosmology. The mathematical nature of the group was discussed originally by Lie (1893) and Campbell (1903). Bateman (1909) investigated the conformal transformations of a space of four dimensions and their applications to geometrical optics, and then worked out the transformations of the electromagnetic equations (Bateman, 1910) along with Cunningham (1910). The results of Cunningham were rediscovered at a later date by Page and Adams (1936a, b); but the new relativity which Page (1936a, b) and Page and Adams (1936a, b) proposed was shown by Robertson (1936a), Engstrom and Zorn (1936) and Bourgin (1936) to be similar to Milne's kinematical relativity and therefore understandable as a special case of Einstein's general theory of relativity. (This is commented on below. Milne's metric for space-time is, in principle, compatible with many theories of gravity, some of which are acceptable and some of which are not. An example of a theory which has cosmological solutions like the Milne model but which is otherwise unacceptable is the Poincaré/Lorentz theory which is discussed by Roxburgh and Tavakol (1975) along with other questions to do with the geometrization of gravity.) After the early work on the conformal group, there was a pause, until it was realized that the group could also be used as a basis to describe geometrical objects (e.g., properties of a space-time). The use of the group in this context was termed generalized projective geometry (Veblen, 1929, 1930). Applied to physics (Veblen and Hoffman, 1930), the resulting field theory was called projective relativity.

The main motivation behind the proposal of projective relativity was to provide a four-dimensional interpretation of the five-dimensional theory of Kaluza (1921) and Klein (1926, 1927). The latter (which is discussed further below) was a metric theory which, at the expense of introducing an extra dimension for space-time, unified the gravitational and electromagnetic fields. The theory of projective geometry in an n-dimensional space was studied in terms of homogeneous coordinates by Van Dantzig (1932), and was applied to physical field theory by Shouten and Van Dantzig (1932). Projective geometry in terms of homogeneous coordinates was also treated by Pauli (1933), who showed that projective relativity was similar to another theory of space-time by Einstein and Mayer (1931). These developments led to a new interest in studying the significance of conformal invariance (Shouten and Haantjes, 1936; Haantjes, 1940; see also Gross and Wess, 1970; Hoyle and Narlikar, 1974, and Chapters 8 and 9), and the role of changes of coordinate frame in classical and special relativistic mechanics (Hill, 1945a), classical electrodynamics (Hill, 1947; Motz, 1953) and general relativity (McVittie, 1942, 1945; Walker, 1945; Infeld, 1945; Infeld and Schild, 1945, 1946; see also Rosen, 1940a, b; Nariai and Ueno, 1960a; Dicke, 1962b; Synge, 1966; Nickerson, 1975; Browne, 1976a; Zel'manov, 1977; Kharbediya, 1977; Roxburgh and Tavakol, 1978; Altschul, 1978). In particular, Hill (1945a, 1947) showed that the introduction of accelerated coordinate systems in relativity is equivalent to going over from the inhomogeneous Lorentz group to the conformal group as a basis for mechanics. That is, instead of the flat-space line element and metric $ds^2 = \eta_{ij} \, dx^i \, dx^j$ of conventional special relativity, one must use the metric of the conformal group (in four dimensions) which is $ds^2 = \beta^2(x)\eta_{ij} \, dx^i \, dx^j$ where β depends on the 15 group parameters (of which 10 form the Lorentz subgroup). The transformations to accelerated axes in classical and relativistic mechanics (Hill, 1945a) yield descriptions of particle motions which are formally the same as those of the Milne and Page theories noted above. To this extent these two theories represent theories of space-time based on the conformal group. They are characterized by a kind of flat-space mechanics (kinematics), in which particles apparently have a new type of kinematical redshift which arises because of the larger nature of the conformal group as opposed to the conventional groups of classical and special-relativistic mechanics (Hill, 1945b). But despite this, one is still only dealing with effects that represent nothing more than a special choice of coordinate frame when considered in the context of the wider theory of general relativity (Robertson, 1936a). In fact the Milne and Page theories represent special choices of the coordinate frame in the Robertson/Walker metric of general relativity (see Robertson, 1933, 1935, 1936a, b, c; Walker, 1935, 1936; Kermack and McCrea, 1933; Milne, 1933b), and interest in the former theories diminished as interest in the latter grew.

Meanwhile, projective relativity had been found by Hoffmann (1947) to provide a natural way of unifying the gravitational field and the vector meson field of particle theory. The general geometry of the conformal group of projective relativity is characterized by invariance under transformations of the Weyl or conformal type, $g_{ij} \to \beta^2(x)g_{ij}$ where β is an unrestricted function of the coordinates. The physically most interesting case of this geometry is where β = constant, since the noted transformation then preserves the shapes of geometrical objects in the space-time (Hoffmann, 1948a). This property suggested to Hoffmann that this special case of the conformal geometry be termed the similarity geometry. He looked at the relation between the gravitational, electrodynamic and vector meson fields in the similarity geometry (Hoffmann, 1948a), and calculated the paths of charged mesons in such a geometry (Hoffmann, 1948b). The status of meson fields and conformal transformations in the wide sense was reviewed by Shouten (1949). A connection between the 15-parameter theory of projective relativity and Dirac's classical theory of electrons was pointed out by Hoffmann (1952). He later broadened the scope of the similarity concept to accommodate an invariance to do with the impossibility of detecting a change in size of both the objects and the standards used to measure them in a space-time possessing the similarity geometry (Hoffmann, 1953a). He then used the new, widened similarity theory of relativity to establish a connection with the Dirac/Schrödinger theory of electrons (Hoffmann, 1953b); but it was some time before conventional general relativity and the Schrödinger theory of elementary particles were connected via the use of the cosmological constant (Nariai and Ueno, 1960b). The connection between Einstein's general relativity and Maxwell's theory of electromagnetism was examined by Gupta (1954), who also gave a review of Einstein's theory in relation to other theories of gravitation that had been proposed as alternatives to it (Gupta, 1957). General relativity as an account of gravitation based on group theory has been considered in the book by Carmeli (1977), while Segal (1977) has re-examined the status of the conformal group in relation to causality and cosmology, restating his belief in the importance of this group for a theory of gravity of the type examined in Chapter 9.

The historical outline just given shows that interest in the conformal group among cosmologists has been persistent, even if it has so far only led to theories, like that of Segal, which cannot be considered to be complete accounts of gravitational phenomena. (The development and shortcomings of conformal relativity have been discussed by Ingraham (1952), who has argued that the conformal group may still provide a valid basis for a metric theory of space-time.) When one considers the course of research into gravity as it has developed in directions other than those to do with general relativity, two things are apparent which may possibly be connected with the reason why so many theories of gravity have shortcomings which eventually lead to their abandonment.

The first thing is the conviction among many physicists that there is something especially attractive about the flat, Minkowski space-time (compare the theories of class 4, Chapter 9). This is mathematically the case, of course, but on a non-anthropomorphic level this preference can only be put into algebraic form by using the concept of conformal flatness. That is, by considering a line element which can be written $ds^2 = \beta^2(x) \, ds_M^2$ where ds_M^2 is the Minkowski line element ($\eta_{ij} \, dx^i \, dx^j$) and $\beta^2(x)$ is a function of the coordinates. This device, which may overlap with metric representations of the conformal group discussed above, is related to the metric foundations of several of the gravitational theories discussed in this chapter and in Chapter 9, and to those of the three theories discussed in Chapter 2. The nature of the assumption of conformal invariance (i.e., invariance under changes in the function β in the general case) has been studied by many workers (see, e.g., Walker, 1945; Synge, 1966; Hoyle and Narlikar, 1974; Bramson, 1974; Bicknell, 1976; Wesson, 1978a; Chapter 9). The power of conformal flatness was realized clearly by Infeld (1945), who noted that the different forms of the Robertson/Walker line element of conventional cosmology can always be written in a conformally Minkowskian manner. This means that the geometrical properties of such a space-time as revealed by light rays must be the same as those of a flat Minkowski metric, although the same is not necessarily true of the gravitational properties of the space-time. Without knowing the details of the latter – i.e., without knowing how the motions of massive bodies are governed by the field equations – one can infer certain properties of the space-time merely from the condition that it be conformally flat. This approach was employed by Infeld and Schild in a pair of fundamental articles in which they isolated a class of cosmological models solely by examining the metric and restricting the form of the function β in accordance with cosmologically desirable properties of the space-time, such as indistinguishability of different places in it one from another (Infeld and Schild, 1945, 1946). It should be emphasized that this procedure is compatible with a range of metric theories of gravity (since it does not depend on the validity of any specific set of field equations). It has wider significance than the less justifiable claims of flatness which typify some of the more modern theories of gravity examined in Chapter 9.

The second thing one notices when one considers the history of gravitation theory is the recurrent theme of embedding conventional (3 + 1)-dimensional models of cosmology in spaces of higher dimensions. The question of whether this is sensible or not depends on the reason for doing it. The Kaluza/Klein theory mentioned above had as a justification for considering a five-dimensional manifold the unification of the gravitational and electromagnetic fields. Nowadays, this might not be considered a strong enough reason by itself for introducing an extra dimension, since electromagnetic phenomena can be included in general relativity by adding the Maxwell energy-momentum tensor for the electromagnetic field to the usual energy-momentum tensor for matter. (Although there may be cases, as considered by Witten (1962), in which algebraic simplicity suggests that a geometrical unification may be in order.) Nevertheless, following Kaluza's proposal of the theory and Klein's development of it (Kaluza, 1921; Klein, 1926, 1927), interest in the theory has continued because it represents a well-established model of a five-dimensional space-time that can be used to evaluate the physical consequences of the assumption that the Universe has an extra dimension. Aspects of the theory were studied by Einstein and Bergmann (1938), Bergmann (1953), Lichnerowicz (1955), Taylor (1962) and Leibowitz and Rosen (1973). It was shown by the last-mentioned authors that the Kaluza/Klein theory can be formulated covariantly for a Riemannian space containing a Killing geodesic vector field. The field equations in the case of empty 5-space correspond to the Einstein/Maxwell equations in 4-space. In the more realistic case of a non-empty 5-space, however, the theory's observable consequences in 4-space are rather bizarre. For example, a 5-space containing dust leads to geodesic equations of motion that correspond in 4-space to the Lorentz equations of motion for particles with arbitrary ratios of charge to mass, and also for tachyons and other kinds of particles that are not observed in the four-dimensional Universe of known physics. These consequences of the five-dimensional Kaluza/Klein theory make one sceptical that it can be brought into conformity with conventional particle physics. There are also some peculiar consequences of the theory as it relates to astrophysics. These latter arise because the field equations for a non-empty 5-space result in an unconventional tensor description of the laws of matter (as they involve energy, momentum, charge and current densities) in 4-space. Most notably, the extra dimension can manifest itself in the four-dimensional world in the form of strange electromagnetic processes in astrophysical systems. This aspect of the Kaluza/Klein theory prompted Taylor (1962) to use it as a technical basis for a suggestion of Bailey that the Universe could be five-dimensional. The data concerned in this suggestion involved planetary and stellar systems, and it is worthwhile to examine briefly the arguments of Bailey and others to see if there is any reason for believing that there are unconventional electromagnetic processes at work in astrophysical systems that indicate a departure from conventional space-time theory.

Bailey (1959, 1960a, b; 1961a, b; 1962) was of the opinion that many astrophysical phenomena (involving cosmic rays and the magnetic fields of the Earth and the Sun) indicate that stars in general carry net negative charges, and that these might arise as a result of the laws of conservation being valid in a five-dimensional continuum rather than in the more familiar four-dimensional one. The size of the charge Q on a star of mass M was suggested by Bailey to be given by $Q = bG^{1/2}M$, where b is a dimensionless constant of size $b \simeq 0.03$ for the Sun so that $Q \simeq 1.5 \times 10^{28}$ e.s.u. for that body. It was pointed out by Motz (1961) that the relation between Q, G and M proposed by Bailey could be connected with a similar relation which gives the gravitational field in the interior of a charged particle, and which can be derived by applying the Weyl theory of gauge invariance to the Ricci curvature tensor in Einstein's theory of gravitation (Motz, 1960). But while the approach of Motz, which will be treated in detail below, may be valid, the relation for stellar charges proposed by Bailey turned out to have unacceptable consequences. In particular, Oster and Philip (1961) showed that the charged-star hypothesis was in contradiction with unambiguous data on the Sun's radiation spectrum and magnetic field which were available at the time, and subsequent research has failed to confirm Bailey's suggestion. The only surviving data which might be compatible with Bailey's hypothesis concern the magnetic fields of rotating stars and other astronomical bodies (like planets) with appreciable angular momenta. That the magnetic momenta of many astronomical bodies are proportional to their angular momenta had been known long before Bailey used some of the data concerned to support the charged-star hypothesis. A combination of theoretical and observational studies of this dependency led to it being termed the Shuster/Wilson effect, and a thorough review of both aspects of the effect was given by Blackett (1947; see also Chapman, 1929, 1948;

Blackett, 1949). However, while recent data on planetary magnetic fields tend weakly to uphold the relationship involved (Russel, 1978), the dependency is not clear-cut and in any case does not necessarily imply any underlying cosmological cause.

A generalization of electrodynamics by Swann (1927) was designed to account for net charges on astronomical bodies by invoking a slow decay of positive charge as opposed to conserved negative charge in cosmic concentrations of matter. This idea was never verified, but appeared in a different form when Lyttleton and Bondi (1959) suggested that a slight difference in magnitude between the electric charges on the proton and electron could give rise to net electric charges on atoms and molecules that, in aggregate, might lead to electric forces powerful enough to produce the observed expansion of the Universe on the basis of Newtonian mechanics. The small net charge expected to be carried by all molecules on the hypothesis of Lyttleton and Bondi was tested for by King (1960), who also reviewed the results of other experiments of a similar nature. The size of the effect was predicted to be related to a proton/electron charge difference of size $|y| \simeq 2 \times 10^{-18}$ e.s.u. when the proton charge is written as $(1 + y)e$; but King's experiment disproved an effect of this size and set a limit of two orders of magnitude less for any uncompensated charge difference in hydrogen and helium molecules. An effect of the Lyttleton/Bondi type could, of course, be produced by inequalities in the *numbers* of protons and electrons (all having charges of identical magnitude) in the Universe, but this sounds *ad hoc* and is difficult to test for. On the other hand, it has more recently been shown by Bally and Harrison (1978; see also Davies, 1978a) that the Universe is in a way electrically polarized because electrons escape more easily than protons from the high-temperature plasma of stellar atmospheres, causing stars to have net positive charges and the interstellar medium to have a net negative charge. However, this latter mechanism, while it produces net charges on stars, is one that involves only conventional plasma astrophysics, and does not necessarily have any connection with the Lyttleton/Bondi charge-inequality cosmology or with the Bailey/Taylor hypothesis on charged stars in a five-dimensional space-time.

One must conclude that neither of these latter hypotheses provide any electrodynamic evidence in favour of a five-dimensional cosmology of the Kaluza/Klein type. This is not in itself a criticism of the use of higher-dimensional manifolds, *provided* these are employed in a passive way as embedding structures for the conventional four-dimensional space-time, rather than as other 'dimensions' producing effects in our space-time that are not in precise accordance with the laws of physics as they are usually formulated. It is known to be mathematically possible to embed homogeneous, isotropic model Universes of Einstein's theory in a flat, five-dimensional manifold, and the embedding problem in general has several mathematically interesting aspects to it (Goldman and Rosen, 1971). One can also in principle make use of embedding spaces of even higher dimension, including infinite-dimensional manifolds (Fischer and Marsden, 1973; see also Kalitzin, 1967). But at some stage one has to ask if the possibility of embedding one space-time in another has any physical (as opposed to mathematical) motivation, possibly in terms of unsolved problems in astrophysics and cosmology. Although there are a lot of the latter, it has to be stated that there is little or no indication that any of them could be solved by the device of embedding a (3 + 1)-dimensional model of cosmology in a space of higher dimensions.

The comments of the preceding paragraphs are not necessarily meant to be critical, although their content in part expresses a negative attitude towards some of the concepts on which theories of gravity can be based. In particular, the lack of a successful theory of gravitation based on the 15-parameter conformal group should not make one sceptical of group theory in general. In the remainder of this chapter a discussion will therefore be given of other aspects of group theory which have connections with gravitation and relativity theory.

10.3. Other Aspects of Group Theory

While conformal relativity is based on the rotation group R_6 of six-dimensional space with its 15 parameters (Ingraham, 1952), the smaller group R_5 gives an extension of special relativity which is again a projective theory of relativity and which has been used by Arcidiacono (1976) as the basis of a cosmology. The new account is akin to Milne's kinematic relativity (see Chapter 2 and Wesson, 1978a), and has the advantage over Segal's theory of accounting for the redshift of the galaxies as a Doppler effect. (A related proposal by Holmberg (1956) that the Universe is describable by a three-dimensional projective space involves accounting for the redshift as a metric effect in a static cosmology and is therefore hardly acceptable in view of the contents of Chapter 11.) On the basis of Arcidiacono's model, the redshift in a model Universe of constant curvature is due to cosmic expansion and is characterized by z-values which are proportional to the recession velocities of the sources. This is encouraging, but the model does not have any metric structure (other than a variant of the flat space-time of special relativity) with which it can accommodate gravitational phenomena.

In this regard, some work of Halpern (1978b; see also Halpern, 1977) is perhaps of more interest since it deals with a formalism for accounts of gravity that have a significant metric structure and are compatible with theories like that of Dirac (Chapter 2), in which importance is attached to dimensionless numbers that are constructed from the parameters of physics. Halpern has examined gravitational theories generated by transformation groups that can be connected with Dirac's Large Numbers Hypothesis. Metric theories closely related to general relativity can be obtained by the requirement that in the local limit the metric be transformable into that of an invariant variety of a group of transformations other than the Poincaré group, in particular the de Sitter group. One of the main justifications for examining metrics compatible with groups like the de Sitter group is that such groups have an established importance in particle physics. Dirac has examined the wave equation in both de Sitter space and conformal space (Dirac, 1935, 1936 respectively). He has formulated field equations for particles like the electron that are covariant with respect to the de Sitter group and the conformal group. Halpern's work is an extension of this approach to see how far gravitation might be expressed in terms of laws which are covariant with respect to the de Sitter group and other groups. (A group theory of gravity by Malin (1974) that uses the de Sitter group as a description of space-time is discussed in Chapter 9.) But while it is of interest to see if theories of gravity that are compatible with the LNH can be formulated in terms of group theory, some problems are present.

Dirac's LNH, which does not allow large dimensionless numbers to be constant in time, is inconsistent with an invariance group with Killing vectors possessing time components (Halpern, 1978b, 1977). Theories which admit of variations of parameters like G are also inconsistent with the Principle of Equivalence, as noted in Chapter 2, although a break with the form of that principle as it is used in Einstein's theory does not necessarily imply a complete abandonment of the equivalence concept. (A partial abandonment of the Principle of Equivalence and the attaching of special importance to cosmic time led Szekeres (1955; see also Kantor and Szekeres, 1956) to a reformulation of general relativity that has some features in common with the theories of Rosen and Katz discussed in Chapter 9. A related reformulation of the field equations of general relativity by Cohn (1975; see also Stephenson and Cohn, 1978a, b) showed that gauge conditions imposed within the formalism of Einstein's theory can also result in departures from the conventional Principle of Equivalence). However, properties such as a departure from the Einsteinian concept of equivalence necessarily mean that a complete gravitational theory that incorporates the LNH will be complex in structure. It is easier to understand the physics of such theories by making use of special atomic and gravitational gauges in the manner of Dirac and others.

Of course, one could take the view that the complex nature of most metric theories of gravity is only a superstructure which provides a dynamical theory for a more fundamental theory in which the physical parameters, such as those used in Dirac's LNH, form the basic content. That is, one could take the view that the values of the constants and other parameters of physics are the prime content of physical theory, with the field equations and their consequences (e.g., equations of motion) being of secondary importance. This philosophy was developed at considerable length by Eddington, whose arguments for taking this kind of approach to physics have been reviewed by Wesson (1978a). Eddington was of the opinion that the parameters of physics ought to be able to be calculated from abstract principles without the need of performing experiments. In particular, he attempted to calculate the values of the fine-structure constant α ($\equiv e^2/\hbar c$ in c.g.s. units and $e^2/4\pi\epsilon_0\hbar c$ in m.k.s. units) and the proton/electron mass ratio m_p/m_e using only algebraic considerations based on the dimensionality of space-time and simple equations with rational coefficients (see, e.g., Eddington, 1936). Among other things, Eddington proposed that the proton/electron mass ratio is given by the ratio of the roots of the quadratic equation $10x^2 - 136x + 1 = 0$, where the presence of the number 136 is connected with the value of α^{-1} ($\simeq 136$). The ratio of roots is approximately 1848. But the observed value of m_p/m_e is closer to 1836. This discrepancy is typical of numerological arguments in physics: they do not work properly, and both Eddington's approach and later modifications of it (see Kilmister, 1966; Atkin and Bastin, 1970; Bastin, 1971; Good, 1970; Sirag, 1977) must be considered unconvincing as far as their observational justification is concerned.

An account which adopts the Eddington philosophy but uses group theory as a basis has been given by Wyler. By considering the structure of the group on which Maxwell's equations (Wyler, 1969) and the potentials for the Coulomb and Yukawa fields (Wyler, 1971) are based, Wyler has claimed to calculate the values of the fine-structure constant and the proton/electron mass ratio from first principles. These two constants are predicted to have the values $\alpha = (9/8\pi^4)(\pi^5/2^45!)^{1/4}$ and $m_p/m_e = 6\pi^5$, which (it must be admitted) look contrived. This is especially true of the first, which is arrived at by calculating volume elements of the seven-dimensional group 0(5, 2). This consists of symmetries expressing five 'real' rotations and two 'unreal' pseudo-rotations, and α is calculated by taking the ratio of the volume elements for the complete seven-dimensional group and the five-dimensional subgroup. The theoretical value of α so arrived at agrees with experiments to within the errors associated with the latter, as discussed by Lubkin (1971).

The theoretical value of $m_p/m_e = 6\pi^5 = 1836.1181$ has been known for a long time to be near to the observed value of m_p/m_e, as first realized by Lenz (1951), and later by Good (1970) who arrived at the noted value by using an extension of Eddington's numerological method. Sirag (1977) has suggested that Eddington's use of algebraic equations as a way of obtaining m_p/m_e be replaced by a combinatorial one based on the number 136 ($\simeq \alpha^{-1}$), this giving $m_p/m_e = 1836$ exactly. But as pointed out by Crawford (1977), neither the exact 1836 value nor the Wyler value of $m_p/m_e = 6\pi^5 = 1836.1181$ agree with the experimental value of $m_p/m_e = 1836.1514$. The first is 21 standard deviations away and the second is 4.6 standard deviations away. Crawford has also drawn attention to the fact that the masses of stable hadrons tend to be spaced at integral multiples of $3m_e$, a tendency which has been examined statistically by Frosch (1973), and which probably has an explanation in terms of elementary particle theory rather than pure mathematics.

The use of abstract concepts to calculate values of the physical constants is, as seen above, not a profitable exercise in terms of the numerical agreement to date of such predictions with observation. From the more subjective side, approaches of this type to accounting for the numerical content of physics are widely felt to be unsatisfying. The reason for this is that most physicists share the conviction that number-fiddling does not by itself constitute valid science in the sense of establishing a causal connection between properties of Nature and the numbers which are used to describe them.

The use of algebra and group theory do not represent the only attempts at calculating the values of the fundamental constants of physics from first principles. A review of other reasonable hypotheses that attempt to reformulate gravitational theory and to evaluate the values of the constants, and especially of G, has been given by Wesson (1978a). Some of these lie only on the edge of being reasonable (e.g., Aspden, 1972; Aspden and Eagles, 1972; Krat and Gerlovin, 1974, 1975a, b; Browne, 1976b). There exist a lot of other accounts which are not reasonable and must be put in the crank class. Disregarding the latter, the hypotheses that attempt to calculate the gravitational parameter G are of two types.

(A) The first type of hypothesis involves establishing a connection between particle physics and gravitation, often via the concept of f (or strong) gravity, which involves the application of general relativity on the elementary particle scale. A concise review of accounts of this type has been given by Fennelly (1974), who has also treated a model in which hadrons are described as tiny Gödell Universes (i.e., as rotating space-times with the equation of state $p = \rho c^2$). The most direct account is that of Motz (1962) who has used the Weyl principle of gauge invariance to construct an enlarged, gauge invariant curvature tensor consisting of the usual Ricci tensor on which Einstein's theory is based plus a term involving the effect of the electromagnetic field. A Lagrangian constructed from this curvature tensor and a metric tensor can be used to study the structure of elementary particles, which on the basis of this model are bodies in which electrostatic repulsive forces are held in equilibrium by a large interior gravitational-type force. (That this might be possible was suggested by the work of Arnowitt et al. (1960), who solved the Coulomb self-energy problem for a point charge by including the gravitational effect of the self-energy.) The force involved can be characterized by a value of G which is much larger than the gravitational constant of conventional Newtonian theory, being

$$G_{\text{strong}} = \frac{\hbar c}{m^2} = \frac{e^2}{\alpha m^2}, \qquad (10.1)$$

where $\hbar$ is Planck's constant divided by 2π, e is the electron charge, m is the mass of an elementary particle and α is the fine-structure constant. With m equal to the proton mass m_p, $G_{\text{strong}} \simeq 2 \times 10^{31}$ gm^{-1} cm^3 sec^{-2} compared to the conventional Newtonian value of $G \simeq 6.67 \times 10^{-8}$ gm^{-1} cm^3 sec^{-2}.

An alternative derivation of (10.1) to that of Motz (1962) has been given by Sakharov (1968). The latter connects up elementary particle theory with the value of G by introducing the concept of the elasticity of space-time. Sakharov's approach, which has been discussed by Zel'dovich and Novikov (1971, pp. 71–74) and Wesson (1978a), has its basis in an identification of the coefficient of the second term in the Einstein field equations with a kind of elasticity of the vacuum. Deriving the equations from an action, S_A, gives the form

$$\frac{\delta S_A}{\delta g^{ij}} = \frac{T_{ij}}{2c} - \frac{c^3}{16\pi G}\left(R_{ij} - \frac{Rg_{ij}}{2}\right) = 0, \qquad (10.2)$$

where the symbols have the usual meanings (g_{ij} is the metric tensor, R_{ij} is the Ricci tensor, R is the curvature invariant, and T_{ij} is the energy-momentum tensor). Equation (10.2) has a form similar to ones in continuum mechanics with an elastic restoring term involving G. The constant of elasticity, $c^3/16\pi G$, is very large, so that the curvature of space produced by an elementary particle of mass m, for example, is very small, the particle being spread over its Compton wavelength h/mc. That is, the curvature produced is very small measured in units of length $h/m_p c$, where m_p is the proton mass. The fundamental gravitational unit of mass, for comparison, is large: $(\hbar c/G)^{1/2} \simeq 2 \times 10^{-5}$ gm (Zel'dovich and Novikov, 1971, p. 66). The object of Sakharov's work is to obtain the elasticity term in Einstein's equations from the curved-space equations of motion as contained in the first term. The T_{ij} also specify the quantum aspects of matter, which are required to connect up with G. This is done by introducing a momentum cutoff (p_0) into quantum theory, such that the curvature correction due to the action S_A gives the elasticity correction as

$$\frac{c^3}{16\pi G} \int R \, dV = \frac{k_A p_0^2}{\hbar} \int R \, dV, \tag{10.3}$$

where k_A is a dimensionless constant of order unity and dV is a volume element. The choice of p_0 as specified by the fundamental mass, 2×10^{-5} gm, gives

$$\frac{m^2 c^2}{\hbar} = \frac{c^3}{G}, \tag{10.4}$$

which is to be read from right to left, and intrinsically defines G in terms of p_0, or the equivalent mass, m. The justification for the choice of p_0 is that it gives approximately the correct value of G, the correct elementary particle charge, and the correct weak-interaction constant.

Equation (10.4) is formally the same as (10.1), but is only satisfied with the Newtonian value of G if one substitutes the mass $m = 2 \times 10^{-5}$ gm mentioned above. Both (10.1) and (10.4) therefore suggest that problems in which Newtonian gravity is dominant with the conventional value of G should be characterized by a fundamental mass of about 2×10^{-5} gm. Zwicky (1961, 1971) examined astrophysical evidence for the presence of a fundamental mass, but on some rather speculative grounds came to the conclusion that if particles exist which deserve to be called 'gravitons' then they ought to have masses of about 5×10^{-64} gm. Although this is almost of the same order as the mass $\hbar H/c^2 \approx 10^{-66}$ gm which one might expect from an application of the Uncertainty Principle to cosmology (Freund et al., 1969; Chapter 9), it is a long way from the expected 2×10^{-5} gm. Markhov (1967) has suggested that typical elementary particle masses ($\approx 10^{-24}$ gm) ought to be derivable from a theory of particle physics based on gravitation by the combination of a dimensionless number of order 10^{-20} with the fundamental mass $(\hbar c/G)^{1/2}$, but such a possibility seems remote in practice. It is likely that if particles with the fundamental mass 2×10^{-5} gm exist and are important in astrophysics, then they are mini black holes that formed in the early Universe with masses corresponding to metric fluctuations on the scale of the Planck length $(G\hbar/c^3)^{1/2} \simeq 1.5 \times 10^{-33}$ cm (Hawking, 1971, 1974; Carr and Hawking, 1974), and not elementary particles in the usual sense.

The Equations (10.1) and (10.4) have a more general significance as can be appreciated by the dimensionalities of the quantities involved. Chandrasekhar (1937), for example, noted that one can construct a series of masses M_d by putting various values of d into the relation

$$M_d = m_p \left(\frac{hc}{G\, m_p^2}\right)^d. \tag{10.5}$$

The choice $d = 3/2$ gives a mass of stellar order, this being also the upper limit to the mass of a degenerate (Fermi/Dirac) astronomical object such as a collapsed star (Chandrasekhar, 1931). The choice $d = 7/4$ gives a mass of the same order as that of the Milky Way; while the choice $d = 2$ gives, roughly speaking, the 'mass of the Universe' (i.e., the mass within the light sphere of size $\approx cH^{-1}$). If one further assumes that $G \propto t^{-1}$, (10.5) gives time-dependencies of the masses of astronomical systems that are similar to those of the Dirac ($\times$)-model considered in Chapter 2.

Another application of (10.1) was pointed out by Motz (1962), who noticed that if $\hbar$ is replaced by the total angular momentum of a system of moment of inertia I and angular velocity ω, then to order of magnitude (10.1) says that $G_{\text{strong}} \approx I\omega c/M^2$ for any mass M. For a system of cosmological size, one can reasonably take $R_U \omega_U = c$ for an idealized model of a rotating Universe with 'radius' R_U. This yields

$$\frac{G\, M_U}{c^2 R_U} \approx 1, \tag{10.6}$$

which is known to apply with the conventional (Newtonian) value of G if the Universe has a density of the order of the closure value $\rho_U \approx 10^{-29}$ gm cm^{-3}.

The account of Motz (1962) outlined above was based on an application of the Weyl theory of gauge invariance to quantum mechanics (Motz, 1960), where it can be used to derive the Bohr/Sommerfeld quantum integral. This approach was extended by Motz (1970) to considering a Reissner/Weyl metric for a charged particle in which there is a discontinuity in the value of G_{strong} at the surface of the particle. The condition $dr/dt = 0$ along the null line $ds = 0$ of the metric (where r is measured from the centre of the particle) gives a quadratic equation whose roots yield the mass of the muon and the mass of the electron. (The Kerr/Newman metric in an f-gravity form, as used in fixing the field at the surface of rotating elementary particles, has been considered by Sexl (1972) and Salam (1972) in the context of high energy particle physics.) Subsequently, Motz suggested that instead of interpreting (10.1) as meaning that the effective gravitational constant inside elementary particles is very large, it would be more sensible to write it as $(\sqrt{G}\, m)^2/c = \hbar$ and treat it as a quantization condition on the square of the gravitational charge $\sqrt{G}\,m$ (Motz, 1972). This condition is also consistent with the Dirac/Schwinger procedure for quantizing charge when that procedure is employed in a rotating Universe model. From the noted quantization condition, Motz deduced that there should exist a fundamental particle in Nature with a mass of the order of that discussed above: namely, the Planck mass $(\hbar c/G)^{1/2} \simeq 2 \times 10^{-5}$ gm. This particle represents, on the basis of the model being considered, a unit that is characteristic of the quantization of mass. Various astrophysical implications of the existence of such particles, including the possibilities of alleviating the missing mass problem in clusters of galaxies and the neutrino discrepancy of solar physics, were discussed by Motz (1972). But as was remarked above it is difficult to find any support for the existence of particles with the noted mass unless they represent mini black holes.

A similar line of argument to that of Motz has been followed by Bakesigaki and Inomata (1971). These authors assume that the S0(4, 2) symmetry of elementary particle theory is broken into the S0(4, 1) symmetry, and derive a formula for the masses of hadrons by assuming that they can be modelled as rotating space-times. (A mass formula for elementary particles had earlier been obtained by Barut and Böhm (1965) using the de Sitter group of space-time as a basis for modelling relativistically rotating particles.) This approach, coupled with the f-gravity theory (Isham et al., 1971), yields a description of elementary particles which is characterized by a value $G_{\text{strong}} \simeq 1 \times 10^{32}$ gm^{-1} cm^3 sec^{-2} for the gravitational constant.

The Motz (1972) approach to the question of how the gravitational field should be quantized serves to reaffirm interest in a topic which has so far been typified by a notable lack of progress in the direction of anything having an obvious physical interpretation. (A short review of the quantization of gravity has been given by Zel'dovich and Novikov (1971, pp. 74–78), while Kuchař (1973) has discussed the geometrodynamical approach to quantization, and Isham et al. (1975) have studied other aspects of quantum gravity.) This would seem to be a field where sophistication does not pay. On the contrary, most insight into the subject of quantized gravity has been derived from the rather crude account of Motz (1972) and earlier formulations of a related problem by G. Rosen (1967) and Harris (1969). The latter workers were concerned with a conjecture that appears to connect up gravitational theory and the sizes of elementary particles. Rosen (1967) considered an integral I over the Ricci curvature scalar density R (which is the Lagrangian density for relativity), thus:

$$I = \frac{c^3}{16\pi G} \int R \, dV. \tag{10.7}$$

The existence of a finite 'size' R_U for the Universe led Rosen to expect that there would be fluctuations in I, these having a characteristic size L of order

$$L \approx \frac{(16\pi G R_U \hbar)^{1/3}}{c} \approx 10^{-12} \text{ cm}, \tag{10.8}$$

which is of the order of distances typical in elementary particle physics. An extension of Rosen's argument by Harris (1969) evaluated the integral over the action of general relativity and put this equal to nh where h is Planck's constant and n is a number. For the case of a particle of radius r situated in a background Universe of 'size' $R_U = cH^{-1}$ the integral can be carried out over a spherically-symmetric field between the limits r and R_U, giving the relation

$$n_1 = -\frac{G m^2}{4rhH}. \tag{10.9}$$

Here, n_1 is the first (negative) quantum mass step, and the relation represents a connection between gravitational theory (characterized by G and H) and elementary particle theory (characterized by h and the particle mass m and particle radius r). The Equation (10.9) holds numerically to order of magnitude with the observed values of G, h and H and values of m and r typical for baryons ($m \approx 10^{-24}$ gm, $r = 10^{-13}$–10^{-12} cm). The result of Harris (1969) combined with Heisenberg's Uncertainty Principle, yields the result of Rosen (1967). Both of these accounts give definite results in (10.9) and (10.8) respectively that hold in the observed Universe and show that it is meaningful to look for a connection between general relativity and elementary particle theory. Indeed, the approach of Harris (1969) can be applied to other metric theories of gravity that can be derived from an action, and Davies (1970) has examined the equivalent problem of the quantization of the gravitational field in the Hoyle/Narlikar theory.

However, the fact that some aspects of the physical world have characteristic lengths (L, r) and masses (m) that satisfy relations like (10.8) and (10.9) does not necessarily mean that these systems are uniquely defined from a quantum mechanical point of view. (For example, different particles can satisfy (10.9) depending on their values of r and m, while it has been seen above that the numerical value of G which appears in both (10.8) and (10.9) need not necessarily be the Newtonian one.) In particular, while the length $L \approx 10^{-12}$ cm can be interpreted as having something to do with elementary particle physics, there is still no accepted theory of particle dynamics and gravitation which yields characteristic lengths in a more direct way than by the rather roundabout method of integrating the particle interaction over a domain of space-time and assuming that the total interaction is quantized. One account which does yield characteristic lengths in a quite direct manner is that of Kursunoglu (1974a; see also Kursunoglu 1952, 1957, 1960, 1974b), in which electric, magnetic and gravitational interactions are described in terms of a theory of gravitation with a non-symmetric metric tensor. (In its original form (Kursunoglu, 1952) the theory was an extension of the Einstein (1951) unified theory of electromagnetism and gravitation.) The theory yields length scales of order 10^{-15} cm and 10^{-25} cm which are suggested as being characteristic of leptonic and hadronic processes respectively; but while in its original form (Kursunoglu, 1952) it might qualify as a viable theory of gravitation and electromagnetism, in its more modern form (Kursunoglu, 1974a) it is characterized by an unlikely metric structure for space-time and an equally unlikely model for the structure of elementary particles based on the doubtful concept of magnetic charge.

While it is not clear if the Kursunoglu theory represents a viable account of particle physics and gravitation, it shares with other metric theories of gravity the property of attaching special significance also to a scale of the order of the Planck length ($\approx 10^{-33}$ cm) mentioned above. But it must be emphasized that this length is not so much typical of elementary particle physics as of quantum gravity processes (see below). The length scales of order 10^{-12}, 10^{-15} and 10^{-25} cm which have been discussed above may have something to do with elementary particle theory, but this is not the same as saying that the metric background used in deriving these lengths is quantized on those scales. In fact, insofar as the length scale $L \approx 10^{-12}$ cm of (10.8) is concerned, this cannot be the case, since Bailey *et al.* (1977) have set a limit on the length scale of any possible quantization of the metric background of the world (e.g., by mini black holes) by measurements of the relativistic time dilation effect for positive and negative muons moving in circular orbits. This experiment, the implications of which have been discussed by Wilkie (1977), tested the time dilation formula of special relativity to within 2 parts in 10^3 (i.e., to 0.2 percent accuracy) for a dilation factor of 30, and found it to be correct at this level. This result, which incidentally upholds the invariance of CPT in weak interactions, leads to the conclusion that space-time is not quantized on length scales down to 10^{-15} cm or so. Ginzburg and Frolov (1977) have discussed the implications of a limit to quantization of this size, and have suggested that it might be improved upon by astrophysical means. It should be appreciated, though, that the noted limit still leaves a long way to go before experiments can begin to detect quantum effects on the scale of the Planck length $(Gh/c^3)^{1/2} \simeq 1.5 \times 10^{-33}$ cm. (A discussion has been given by

De Witt (1962) of how geometry can formally be quantized and of the length scales one might expect to figure in such a quantization.) On dimensional and other grounds one might expect quantum fluctuations of the metric background of space to be of order unity on scales of the size of the Planck length (Ginzburg et al., 1971), but the detection of such effects must await the development of new experimental techniques.

(B) The second type of hypothesis that has been used to find a value for G (or at least, 'explain' its size) involves the Eddington numbers. These latter are dimensionless numbers characteristic of astrophysical and cosmological data that are usually multiples of 10^{40}. A comprehensive review of the sizes and possible explanations of these numbers has been given in Wesson (1978a). Broadly speaking, accounts of the nature of these numbers depend either on elementary particle theory or on cosmology (in the latter context, equalities between Eddington numbers composed of constants of physics from different disciplines are often termed the cosmological coincidences). The equality expressed by Dirac's Large Numbers Hypothesis in the form of Equation (2.1) of Chapter 2 is a way of accounting for the sizes of the Eddington numbers by using G-variability. The dimensionless numbers M_d/m_p obtained from Equation (10.5) for some values of d give numbers which are powers of 10^{40}; and if one could find any justification from astrophysics for choices of values of d that yield only powers of 10^{40} in (10.5) then one could say that these Eddington numbers have an astrophysical explanation. The problem with both the LNH and the formal relation (10.5) is, of course, that they do not 'explain' the Eddington numbers in the usual sense of the word but rather quantify the relations between Eddington numbers of various sizes. In this respect, both Dirac's formula (2.1) and Chandrasekhar's formula (10.5) suffer from the drawback met with above in the discussion of group-theoretical bases for the values of the parameters of physics: they provide evaluations, not explanations.

Some explanations which deserve the appellation have been reviewed in Wesson (1978a), but none has been universally accepted, and there is a general feeling among cosmologists that a scheme which incorporates an account of the Eddington numbers of all sizes has yet to be found. One should also realize that new distinctive numbers and cosmological coincidences are likely to turn up as research into astrophysics progresses. Some puzzling coincidences which have physical aspects in common but which involve such diverse subjects as the expansion of the Earth and the Hubble velocities of the galaxies are already known (Wesson, 1973, 1975a); but no reliable explanation of these data is available. A brief discussion of the data concerned and a tentative explanation of them based on scale invariance has been given in

Chapter 6, based on work by Wesson (1973, 1975a, 1978a). A more speculative restatement of the same idea and the data concerned has been made by Pavšić (1975) and Selak (1978). However, it must be emphasized that the main thing that these data do is to indicate that there are numerical coincidences present in our descriptions of astrophysical processes that cannot be explained satisfactorily with conventional theory. This should make one wary of constructing 'closed' accounts of the cosmological coincidences and the Eddington numbers based on standard physics. This does not mean to say, though, that certain of the numbers involved cannot be understood on the basis of hypotheses that might later prove to be parts of some wider theory.

For example, Silk (1977a, b, c) has studied the physics of gas dynamical processes, of the type believed to be important in the formation of astrophysical systems such as galaxies and stars by the collapse and fragmentation of matter under the influence of gravity. He has used these calculations (Silk, 1977d) to provide an astrophysical explanation for the size of the dimensionless parameter $Gm_p^2/hc \approx 10^{-40}$ ($d = 1$ in (10.5)). This account of the size of a number which has often been discussed in a purely numerical way as a cosmological coincidence thus explains the size of it in terms of the existence of galaxies and stars. The existence of these systems is actually a prerequisite for the presence in the Universe of observers like human beings. This suggests that the presence of life in the Universe may not be something which is totally independent of the values of the constants of physics. This idea has been quantified in one form by Clutton-Brock (1977), who has examined the value of the entropy per baryon in the 'many-worlds' cosmology in which it is imagined that many Universes exist simultaneously having different values of the constants of physics (Everett, 1957; De Witt and Graham, 1973). The entropy per baryon in our Universe is mainly represented by the photons of the 3 K background and the protons in the galaxies (see Chapter 3 and Chapter 11), and is of order of magnitude 10^9. Clutton-Brock has shown that life as we know it is only consistent with model Universes having values of the entropy per baryon in the range $3 \times 10^5 - 5 \times 10^{11}$. In those with values less than the noted lower limit, only metal-poor dwarf galaxies form in which no planets evolve to support observers of human type; while in those with values greater than the noted upper limit, galaxies do not form at all, but rather massive black holes instead (Clutton-Brock, 1977; see also Clutton-Brock, 1976). These considerations suggest that the value 10^9 for the entropy per baryon in our Universe may not be accidental. The work of Silk and Clutton-Brock is representative of similar accounts (see Wesson, 1978a) which connect up the sizes of the Eddington numbers and the origin of the cosmological coincidences with astrophysical conditions necessary for the existence of human-type observers.

Astrophysical conditions have also been used by Barker (1977), who has drawn attention to a coincidence in length scales – between data to do with the expansion of the Universe and the energy density of the 3 K microwave background – which is similar in nature to the much-studied cosmological coincidences between the Eddington numbers. Barker has noticed that the radius of the Hubble sphere ct_0 is approximately equal to the mean free path of cosmic ray protons propagating through the photons forming the universal 3 K radiation field. The quantity ct_0 defines the size of the light sphere or horizon in an expanding, big-bang Universe of age t_0, and is mainly of dynamical significance since $t_0 \simeq H^{-1}$. Contrarily, the mean free path of a proton propagating through a photon field with number density n_{ph} is $(\sigma n_{ph})^{-1}$ where σ is the Thompson cross-section for the proton ($\sigma = 8\pi e^4/3m_p^2 c^4 \simeq 2 \times 10^{-31}$ cm^2), and so depends on particle parameters (e, m_p, c) and the 'temperature of the Universe' ($T \simeq 3$ K). The former length is $ct_0 \simeq 1\text{--}2 \times 10^{10}$ light years ($\simeq 1\text{--}2 \times 10^{28}$ cm). The latter length is $(\sigma n_{ph})^{-1} \approx 10^{28}$ cm to order of magnitude, where $n_{ph} \approx 10^2$ cm^{-3} (Greisen, 1966; Barker, 1977). This coincidence holds numerically with the noted values of the parameters involved. But whether or not it has the physical interpretation that the length $(\sigma n_{ph})^{-1}$ actually is a cosmic ray/3 K background mean free path for scattering depends on the physics of the scattering processes involved and on where cosmic rays come from. The interaction of cosmic rays with the microwave background has been extensively studied in relation to the question of the origin of cosmic rays (Hoyle, 1965; Greisen, 1966; Felten and Morrison, 1966; Wdowczyk and Wolfendale, 1975). In particular, the mean free path of cosmic rays originating at cosmological distances is in practice probably fixed by pair creation and photo-pion production (Greisen, 1966) rather than by Compton scattering (Hoyle, 1965; Barker, 1977). The former processes, if dominant, predict a mean free path of order 10^{25} cm and a sharp cutoff in the spectrum of cosmic rays at energies of order 10^{20} eV due to energy loss by particle production during scattering of high energy ($>10^{20}$ eV) cosmic rays off the cold (3 K) microwave photons. Although the data involved are not known with great accuracy, there are already some observations that seem to imply that the cutoff is not present at the expected energy (see Chapter 11 for a further discussion of this topic in relation to the question of the universality of the 3 K background). The absence of the expected cutoff would mean either that the cosmic ray mean free path is not fixed by photo-meson production, or that the cosmic rays do not originate at cosmological distances, or that the microwave background is not all-pervasive. In view of these uncertainties, it is not clear that Barker's coincidence has a physical meaning in terms of cosmic ray/photon interactions; but as noted above, it does have a numerical meaning insofar as it involves the cosmological parameters t_0 (or H) and c and the particle parameters e, m_p and c out of which the cross-section σ is formed.

The comments of the previous four paragraphs show that the status of the Eddington numbers is equivocal, and that in practice they do not explain properly the size of G or of the other constants of physics. The status of the preceding methods of class (A) is perhaps better in this respect, but while strong gravity can provide a connection between elementary particle theory and gravitation, a complete theory which encompasses both fields is still lacking. Despite the failure hitherto experienced in finding an acceptable theory of gravity that is based on the conformal group, it could well be that group theory in some form will eventually provide the basis for a properly integrated theory of elementary particles and gravitation.

CHAPTER 11

THE STATUS OF NON-DOPPLER REDSHIFTS IN ASTROPHYSICS

11.1. Introduction

It was mentioned elsewhere that there are certain theories of gravitation (most notably those of Hoyle/Narlikar and Segal, discussed in Chapter 2 and Chapter 9 respectively) which are connected with the existence of redshifts that do not arise from the Doppler effect acting in an expanding general-relativistic Universe. The purpose of this chapter is to take a look at possible evidence for non-Doppler redshifts in astrophysics to see if there is any case for believing that astronomical systems provide a basis for departing from conventional big-bang cosmology as founded on Einstein's general theory of relativity.

Arguments which are relevant to this question can be divided up roughly into three categories. At a basic level, some workers find the idea of an expanding Universe disagreeable and reject the expansion postulate (for various and often unexplained reasons), replacing the Doppler mechanism for producing redshifts by an alternative one. The status of this argument will be examined first below. Then there are various claims of the existence of redshift anomalies which, while not necessarily incompatible with the expanding Universe postulate, nevertheless imply the operation of an unconventional redshift-producing mechanism in some astronomical sources. The status of this argument will be examined second below. Lastly, there is the possibility that there exist trends in the redshifts of astronomical sources which, while anomalous, are of conventional origin (meaning that the redshifts involved, unlike those of the two previous categories, are produced by processes compatible with known Einsteinian physics). Such anomalous redshifts are only anomalous in the sense of being incompatible with the simple Friedmann models of the Universe.

11.2. The Tired-Light and Related Theories

The tired-light theory is based on the idea that the wavelength of the light from galaxies is redshifted because of changes in the photons which occur during the long journey from source to receiver. It is perhaps the most commonly advocated non-Einsteinian alternative for accounting for the existence of the cosmological redshift. There is no direct laboratory evidence for photon properties of the type invoked by tired-light theories, although a laboratory experiment to test for a possible ageing of photons was proposed by Shamir and Fox (1967); but various phenomenological relations for the redshift involved have been discussed, some of which are based on theoretical as well as observational considerations (see below). For redshifts produced as a result of energy loss in photons travelling for long periods t through intergalactic space, a common formula for the value of the redshift is $1 + z = \exp(Ht)$, where Hubble's parameter H is taken as a given constant (Gerasim, 1969; Horedt, 1973). The use of a constant like H here is something which needs to be justified if it is employed in a non-expanding cosmological model, but of course the main thing needed to justify such models is a plausible mechanism for producing non-Doppler redshifts.

One much-discussed mechanism was due in original form to Finlay-Freundlich (1954a, b) who suggested that photons passing through a radiation field might lose energy in proportion to the path length and the fourth power of the temperature of the field. This relation was an empirical one, based primarily on studies of stellar redshifts, but it was also used to estimate the temperature (i.e., energy density) of the intergalactic radiation field. When photons lose energy, their wavelengths increase and they exhibit a redshift, so in principle a mechanism of the type proposed by Finlay-Freundlich could account for the redshifts of the galaxies if photons on their way from distant sources interact with an intergalactic radiation field. One such radiation field (that of the 3 K microwave background) is known to exist, and there might also be a notable intergalactic radiation field at optical wavelengths if an argument of Ward (1961) is accepted. The latter involved a balance between the rate at which light is released into intergalactic space from stars and the rate at which it is degraded in energy by the redshift, the balance providing a radiation field of the type required by Finlay-Freundlich's redshift mechanism. (A similar application of the argument that redshifts arise when photons pass through an intergalactic radiation field will be mentioned in a discussion of the Rubin/Ford/Rubin redshift anomaly below.) This question has ramifications in other areas of cosmology, so it is instructive to take a critical look at the argument involved using modern data.

If a typical galaxy consists of 10^{11} stars each radiating (mostly optical-band) energy at a rate of about 4×10^{33} erg sec^{-1}, and if the average distance between galaxies is taken as about 300 kpc, then the rate at which starlight energy is being pumped into intergalactic space per unit volume of the latter is $R_+ \simeq 1 \times 10^{-28}$ erg cm^{-3} sec^{-1}. If the energy density of the field is ρ_r, the rate at which it is being degraded is $R_- = \rho_r/T$, where T is a time scale which is of crucial importance in fixing ρ_r and which is to be evaluated on astrophysical data (see below). The condition that the intergalactic radiation field should maintain its equilibrium density of ρ_r is simply that $R_+ = R_-$ for a steady-state balance. Some comments may now be made as regards the existence and energy density of such a field and Ward's argument in favour of the Finlay-Freundlich type of mechanism.

Firstly, it is logically wrong to assume that the time scale T is H^{-1} as taken by Ward, since Hubble's parameter is connected with the hypothesis of the expansion of the Universe, and so cannot be used in an argument for radiation-produced redshifts without incurring an error in reasoning.

Secondly, if one uses for T a typical galactic age (i.e., $T \simeq t_0 \simeq 10^{10}$ yr), one is faced on the non-expansion basis with explaining the coincidence that the redshifts of the galaxies vary with distance according to the law $z = Hr/c$ (Chapter 9) where H is a constant with the dimensions of an inverse time that just *happens* to have a value such that $H^{-1} \simeq t_0 \simeq 10^{10}$ yr. This seems rather fortuitous if the Universe is not expanding, since there is no reason why the H in the redshift law should be equal to the reciprocal of a typical galactic age if the redshift is not due to expansion. A collateral argument in favour of the conventional interpretation of the size of $T \simeq t_0$ is that Olber's 'paradox' is in practice avoided by the fact that stars do not shine long enough in comparison to the time scale $T \approx 10^{10}$ yr to fill up space with their radiation (Harrison, 1977). Tired-light theories in which the Universe did not begin in a big bang have, contrarily, no natural explanation of why the night sky is as dark as it is. This is discussed quantitatively in the next paragraph.

The third objection to the Ward argument in favour of the Finlay-Freundlich mechanism is that if, for the sake of argument, one does take $T \simeq 10^{10}$ yr $\simeq H^{-1}$, then the energy density ρ_r would be expected to be $\rho_r \approx R_+ T \simeq 3 \times 10^{-11}$ erg cm^{-3}. This is unacceptably large. It is two orders of magnitude larger than the energy density of the 3 K microwave background (Wesson, 1975b, 1977b; Wesson and Lermann, 1976; see below). Furthermore, it is in contradiction to studies of the extragalactic light from both the theoretical side (Whitrow and Yallop, 1964, 1965; Bonner, 1964; Partridge and Peebles, 1967a, b; Peebles, 1971a, pp. 59–63, 1971b; Pegg, 1971; Shectman, 1973; Hara, 1974; Tinsley, 1977b, 1978b, c) and the observational side (Roach and Smith, 1965; Lillie, 1972; Shectman, 1974; Matilla, 1976; Dube *et al.*, 1977; Hofmann and Lemke, 1978; Spinrad and Stone, 1978; see also Maucherat-Joubert *et al.*, 1978). Of the observational studies, only that of Matilla (1976) gave a finite result (namely that the intensity was 23 ($\pm$8) $\times 10^{-9}$ erg cm^{-2} sec^{-1} ster^{-1} Å^{-1} at a wavelength of

4000 Å; this may be converted to an approximate energy density by multiplying by 4π(ster) and 4000 (Å) and dividing by $c = 3 \times 10^{10}$ (cm sec^{-1}), giving $\rho_r \simeq 3\ (\pm 1) \times 10^{-14}$ erg cm^{-3}). But this result is in conflict with that of Dube et al. (1977), who obtained a quite stringent upper limit on the extragalactic background light (of $1.0(\pm 1.2)$ S_{10} units, where 1 S_{10} unit $\simeq 7.5 \times 10^{-6}$ erg sec^{-1} ster^{-1} at a wavelength of 5100 Å; this restricts ρ_r to $\rho_r \lesssim 4 \times 10^{-15}$ erg cm^{-3} at a wavelength of 5100 Å as used by Dube et al.). It is also in conflict with an upper limit obtained by Spinrad and Stone using Mattila's method (this limit being 4×10^{-20} erg cm^{-2} sec^{-1} ster^{-1} Hz^{-1}, corresponding to an upper limit on the energy density of $\rho_r \lesssim 1.2 \times 10^{-14}$ erg cm^{-3} at a wavelength of 4000 Å as used by Spinrad and Stone). These data show that the intergalactic radiation field has an energy density not larger than $\rho_r \approx 10^{-14}$ erg cm^{-3} at optical wavelengths. (It might be larger at other wavelengths where there are as yet sparse data available with which to set a limit. Hofmann and Lemke (1978) set an upper limit to the inensity of 6×10^{-11} W cm^{-2} ster^{-1} μm^{-1} at a wavelength of 2.4 μm = 24 000 Å, corresponding to $\rho_r \lesssim 6 \times 10^{-13}$ erg cm^{-3} at this wavelength.) These data on ρ_r may be compared to the energy density of the 3 K microwave background, which is about 4×10^{-13} erg cm^{-3} and which happens to be comparable to other energy densities connected with astrophysical processes within the Galaxy (Hoyle et al., 1968; Peebles, 1971a; Wesson, 1975b). One thus sees that any intergalactic radiation field in the optical band has an energy density which is at least one order, and probably two orders, of magnitude less than the energy density of the 3 K background at centimetre wavelengths (see also Blair, 1974). The optical field is therefore of negligible importance compared to the microwave one in the intergalactic case. Further, there is no evidence that the 3 K field itself can produce redshifts in optical photons moving through it, and Steigman (1978) has shown that tired-light theories are incompatible with the observed Planckian spectrum of the microwave photons which comprise the 3 K background.

The preceding three comments tend to argue strongly against Ward's suggestion of an energy-balance basis for the Finlay-Freundlich redshift mechanism (Ward, 1961; Finlay-Freundlich, 1954a, b) and against the existence of conditions in general in intergalactic space that could lead to a redshift caused by an optical radiation field. Although it is possible to provide a theoretical basis for the Finlay-Freundlich type of tired-light mechanism in terms of otherwise unknown photon-photon interactions (Born, 1954; Ter Haar, 1954a), there are, in addition to the three comments given above, numerous other objections that can be brought against radiation-produced redshifts (McCrea, 1954; Burbidge and Burbidge, 1954; Ter Haar, 1954b;

Helfer, 1954; Melvin, 1955). The direct test of the Finlay-Freundlich mechanism that was proposed by Ward (1961) does not seem to have been carried out, and so in summary one can say that there is no laboratory evidence in favour of it and a lot of astrophysical evidence against it.

Despite this, tired-light mechanisms for producing the redshift continue to be discussed. One such hypothesis has been proposed by Pecker et al. (1972), who have invoked a finite rest mass for the photon in order to account for apparent non-velocity redshifts of the type which form the astrophysical motivation for the Hoyle/Narlikar cosmology. The photon rest mass must necessarily be less than 10^{-49} gm or so by the results of Franken and Ampulski (1971) and Williams et al. (1971). There are also other problems with mechanisms of the type proposed by Pecker et al. (1972), and their hypothesis has been criticized by Woodward and Yourgrau (1973), Aldrovandi et al. (1973) and Chastel (1976), whose arguments it is worthwhile to examine.

Woodward and Yourgrau (1973) have given a critical appraisal of the hypothesis of Pecker et al., and have discussed in particular the possible relation between photon-photon interactions and evidence for anomalies in the propagation of electromagnetic waves passing near massive bodies such as the Sun. Several different types of anomaly have been reported (see below), and interpreted in different ways. For example, following Pecker et al. (1972), Merat et al. (1974) have attributed a freak wavelength shift in the signal from the Pioneer-6 spacecraft in the vicinity of the Sun to an anomalous interaction between photons and neutral bosons of small mass that might have a solar origin. This type of phenomenon is compatible with the hypothesis of Pecker et al., but Woodward and Yourgrau have argued that while a tired-light mechanism of the type involving a finite rest mass for the photon may be partially valid and help to explain some astrophysical problems (such as that of the virial discrepancy in clusters of galaxies; see elsewhere), it is not adequate as an explanation of the darkness of the night sky and of evidence in favour of other types of anomaly in the propagation properties of electromagnetic waves in the Solar System.

Woodward and Yourgrau (1973) believe that the latter especially indicate an anomalous photon-graviton (as opposed to photon-photon) interaction, but in fairness to the tired-light hypothesis it must be stated that the data involved are themselves somewhat contradictory and permit of several different interpretations. In order to gain an overview of the data concerned, it is useful to list here the main results which have been discussed over the years as they relate to possible photon-photon and/or photon-graviton effects of non-Einsteinian origin. (i) Wheelon (1952) accounted for an apparently non-Einsteinian contribution to the deflection of light by the Sun by invoking a finite rest mass for the photon; but more modern data noted in Wesson (1978a) show that there is no departure from general relativity that can be used to justify a significant photon rest mass as far as the three classical Solar System tests of that theory are concerned. Neither does the fourth test, involving the time delay of an electromagnetic wave passing near the Sun (Shapiro et al., 1968) give any grounds for invoking a finite photon rest mass. (ii) Sadeh et al. (1968a) observed an anomalous frequency shift in the radiation coming from a remote astronomical source as that signal passed near the Sun. This anomaly was discussed in terms of a non-Einsteinian effect of mass on the frequency of electromagnetic waves by Sadeh et al. (1968b), who also carried out an Earth-based experiment that showed a similar anomalous dependency of frequency on the gravitational field. These experiments caused much comment. Szekeres (1968) viewed the effect involved as observational support for a spinor field theory with a non-Riemannian connection which he had developed earlier (Szekeres, 1957). But a laser experiment by Shamir (1969) failed to confirm an expected effect of the Szekeres theory as the consequences of that theory could be worked out using data from the experiments of Sadeh et al. (1968a, b), and other explanations of the results of those experiments were quick to follow. Woodward and Yourgrau (1970a) interpreted the effect as an anomalous photon-graviton interaction of a type in which the velocity of light was supposed to depend on its frequency. This hypothesis, it was later realized, was actually in conflict with the data concerned in the form in which the hypothesis was first proposed (Reinhardt, 1971; Woodward and Yourgrau, 1971). An anomalous photon-graviton interaction as an explanation of the results of Sadeh et al. (1968a, b) was also criticized by Ferencz and Tarcsai (1971), who made a case for believing that the anomaly involved could be understood in terms of conventional processes of solar physics. The question of anomalous interactions between electromagnetism and gravitation was re-discussed by Woodward and Yourgrau (1972), along with the results of experiments relevant to the existence of propagation anomalies in the Solar System; but while such propagation effects might be explicable on the basis of photon-photon or photon-graviton interactions, the need for such mechanisms was vitiated by the results of a series of experiments which had been prompted by the work of Sadeh et al. (1968a, b). (iii) Sadeh et al. (1968c) themselves carried out an experiment on the time of arrival of pulsar radiation, but failed to observe an effect expected on the basis of their earlier experiment (Sadeh et al., 1968a). Likewise, Ball et al. (1970) failed to confirm the effect observed by Sadeh et al. (1968a). The Earth-based experiment of Sadeh et al. (1968b), which had shown the presence of an anomalous frequency shift for electromagnetic waves travelling long distances in the Earth's gravitational field, was also not confirmed, since an experiment by Markowitz (1968) failed to show any anomaly. (iv) A related class of experiments which were carried out at about the same time showed that the velocity did not depend appreciably on the frequency (or wavelength, equivalently) in light coming from remote astronomical sources (Warner and Nather, 1969; Isaak, 1969; Feinberg, 1969). This conclusion was discussed by Woodward and Yourgrau (1970b), but as pointed out by Synge (1969) questions such as this depend as regards their interpretation on how one defines the parameters involved and on what standard of reference for space-time one refers them to. This latitude of interpretation does not alter the fact that the noted experiments failed to show any non-Einsteinian anomaly. (v) It was mentioned above that Merat *et al.* (1974) have interpreted an anomaly in the signal broadcast by the Pioneer-6 spacecraft as support for a tired-light theory of the type proposed by Pecker *et al.* (1972). However, while observations of Pioneer-6 at conjunction did reveal anomalies in signal propagation, these appear to have been only transient, with no persistent effects that could not be explained by processes of conventional physics (Levy *et al.*, 1969; Goldstein, 1969; see also Ferencz and Tarcsai, 1971). This last remark brings one back to the question of whether there is any evidence of anomalous photon-photon or photon-graviton interactions in data to do with the propagation of electromagnetic waves in the Solar System. Quite apart from the apparent inadequacy of tired-light mechanisms to account for such anomalies if they exist (Woodward and Yourgrau, 1973), one sees that in practice there may well be nothing for the tired-light hypothesis of Pecker *et al.* (1972) to explain in the first place.

In addition to the negative nature of the preceding remarks, the hypothesis of Pecker *et al.* (1972) has been subjected by Aldrovandi *et al.* (1973) to a criticism which can be simply stated but which is of great importance for this and other photon-photon tired-light mechanisms. Aldrovandi *et al.* have argued that the photon-photon mechanism would produce smeared-out grey radiation, rather than a shifted discrete spectral line, in the light coming from a source affected by such a mechanism. This criticism appears to be well founded, and while Pecker *et al.* (1973) have defended the feasibility of a redshift-producing mechanism in quasars and other astrophysical sources that depends on a finite rest mass for the photon, one obtains the impression that such a mechanism is not viable.

The non-viability of photon-photon redshift mechanisms has been underlined by Chastel (1976), whose comments represent a third and cogent criticism of the hypothesis of Pecker *et al.* (1972). It has been argued by

Chastel that photon-photon interactions of types that could produce significant astrophysical effects would lead to notable contradictions with quantum electrodynamics, which theory is well established by laboratory experiments. In addition to this, Chastel has concluded that even if a photon-photon interaction of the type proposed by Pecker *et al.* exists, then it cannot explain anomalous redshifts in single and binary stars, QSOs and galaxy groups of the type studied by Arp (see Section 11.3). The points which have been brought out by Chastel also make it unlikely that the interaction of optical photons with the microwave photons of the 3 K background can represent a viable tired-light mechanism for the origin of the redshifts of galaxies.

A tired-light mechanism for producing a redshift that is similar in type to that considered by Pecker *et al.* (1972, 1973) has also been proposed by Weinstein and Keeney (1973a, 1974). This mechanism, which also involves a finite rest mass for the photon but which has not been applied in detail to explaining the redshift anomalies in astrophysical sources, has been discussed at length by Wesson (1978a) and might be viable as judged from an astrophysical standpoint. However, unlike other tired-light theories, the work of Weinstein and Keeney was motivated by a wish to explain some anomalous trends in the long-term behaviour of the Earth–Moon system (Weinstein and Keeney, 1973b, 1975; Wesson, 1978a). It is characteristic of fundamental processes such as those involved in tired-light cosmologies that they affect physical systems of many different types. There are, though, no firm data in support of such theories from geophysics and planetary physics (Wesson, 1978a), and the only notable argument in favour of them concerns possible non-Doppler redshifts in astrophysics.

In this respect, Vigier (1977) has reviewed the evidence for the existence of non-velocity redshifts, particularly as these might be connected with the tired-light hypothesis. There are several different types of argument which have been made for the presence of anomalous redshifts in quasars and some galaxies (see below), but even if such exist they do not necessarily indicate a tired-light redshift-producing mechanism. There is also some evidence for non-velocity redshifts in observations of stars, and Kuhi *et al.* (1974) have examined the status of stellar redshift-producing mechanisms, especially in relation to the anomalous redshifts of order 100 km sec^{-1} which are observed in the emission line spectra of Wolf/Rayet binary star systems. However, they conclude in this case that while the anomalous z-values could be due to inelastic photon/boson scattering, they could also be due to conventional absorption processes in moving matter. There is certainly no proof of the operation of processes which depend on a finite photon rest mass in stellar systems. This renders implausible any tired-light mechanism that depends on photon interactions, and since it is unlikely that such processes could operate on extragalactic scales without some evidence of them being found on smaller scales, one must conclude that there is no basis for tired-light cosmologies based on as yet unverified properties of photons. This conclusion has been re-emphasized by Geller and Peebles (1972; see also Peebles, 1971a). They have examined the status of cosmological observations as they relate to the tired-light idea, and have found that this hypothesis is much inferior to the conventional expansion hypothesis in accounting for the observed redshifts of the galaxies and other astronomical objects.

Some time later, Ellis (1978b) attempted to show that the question of whether the Universe is expanding or not could be answered in the negative. Unfortunately, this involved a picture of the Universe in which we as observers are supposed to be living near one of two singular poles in a static, inhomogeneous world model, and in which agreement with observations of redshifted galaxies is only obtained as a bizarre contrivance which most cosmologists would be extremely reluctant to accept. A review of the static Ellis model was given by Davies (1978b); but in a subsequent treatment (Ellis *et al.*, 1978) it was realized that a static model of the type considered by Ellis cannot yield the observed magnitude/redshift relation for galaxies. This result indicates directly that the m/z relation for real galaxies depends on the fact that the Universe is expanding.

11.3. Redshift Anomalies in Astronomical Sources

Without necessarily going outside the expanding Universe picture, there have been several claims that the redshifts of some astronomical sources show anomalies which cannot be explained by conventional models of the Universe based on Einstein's theory of general relativity.

(a) Arp has for long claimed that QSOs and some galaxies are physically associated, despite the widely discrepant redshifts of the members of any given pair or group of objects. (Reviews of the evidence for non-velocity redshifts have been given by Arp (1971a, 1974a, b) along with discussions of particular cases which favour their existence.) One special type of association that has been much discussed concerns those cases in which objects are aligned with regular spacings but display different redshifts, a situation that might arise if secondary objects are periodically ejected from a parent galaxy. However, an alternative explanation for sources which seem to be regularly spaced but which are really at different distances from an observer has been given by Wesson (1978g) in terms of the presence of dark intergalactic matter, possibly in the form of haloes around galaxies or clusters of galaxies. The implications of claims like that of Arp affect many fields of astrophysics (see the book by Field *et al.*, 1973), and finding an explanation for redshifts that did not seem to be due to the Hubble expansion was one of the motivations behind the Hoyle/Narlikar theory discussed in Chapter 2. This theory may still have some astrophysical basis, since Bottinelli and Gougenheim (1973) confirmed part of Arp's case in showing that companion galaxies have redshifts that are on average 90 km sec^{-1} greater than those of the bright parent galaxies. But some redshift discrepancies of this type might be due to astrophysical processes to do with gas flow, as discussed by Lewis (1975), who has also reviewed evidence for anomalous redshifts of the Arp type and failed to confirm their existence in some systems such as groups of galaxies (Lewis, 1971; Arp, 1970, 1971b; see also Lewis, 1969; Arp, 1966; Arp and Madore, 1975). The matter is still not settled, though. Galaxies which are close together on the sky but show discrepant redshifts, and similar examples of what appear to be galaxy/galaxy and galaxy/QSO pairs with discrepant redshifts, have been studied by Nottale and Moles (1978) and Moles and Nottale (1978) respectively, with the conclusion that such associations are probably *not* due to chance projection. However, the argument on which this conclusion is based is statistical, and many astronomers rightly view with suspicion arguments of this type which are applied to particular cases of pairs or groups of objects with discrepant redshifts (see, e.g., the comments of Wampler *et al.* (1973) and Hazard *et al.* (1973) on QSO pairing). An analysis using a large number of data by Plagemann (1973) has shown that, irrespective of the nature of QSO redshifts, there is no statistical evidence that QSOs are preferentially associated with bright galaxies. This conclusion has been confirmed more recently by Nieto (1977, 1978). Another recent result which goes a long way towards refuting claims, based solely on statistical arguments, of physical association between objects with different z-values is due to Stockton (1978). He has found that QSO z-values are in some cases approximately the *same* as the z-values of galaxies near to them. This proves that at least some QSOs and galaxies which are close together have similar (Hubble-derived) redshifts. Stockton's result strongly suggests that occasional pairs which show discrepant z-values are merely the results of accidental surface projections of objects which are in reality at different distances.

(b) Tifft has claimed that studies of the z-values of some samples of galaxies show the existence of redshift 'bands' (Tifft, 1973). That is, the z-values are not randomly distributed but instead show periodicities, this tendency being possibly connected with fine structure within the magnitude/redshift correlation for galaxies (Tifft, 1974). One apparently dominant z periodicity corresponds to a velocity periodicity of 72.5 km sec^{-1}. This might, in principle, be explained in terms of a mechanism in spiral galaxies which involves the ejection from a common centre of two opposed and outward-flowing streams of material (Tifft, 1976, 1977; see also Lewis, 1975; Freeman, 1965). In practice, this mechanism does not seem to be viable because observations of the galaxy NGC 628 by Monnet and Deharveng (1978) have failed to reveal intrinsic redshift effects of the sort expected on the expanding-arm hypothesis. The question of expansion in spiral arms has also been discussed by Jaakkola *et al.* (1978), although they view evidence for systematic redshift gradients across galactic disks of neutral hydrogen mainly as support for the existence of non-velocity redshifts in galaxies (Jaakkola *et al.*, 1975a). But even if one grants that there exist redshift anomalies in spiral galaxies that cannot be accounted for by observational biases, the non-velocity hypothesis represents a more contrived and therefore less likely explanation of them than does arm expansion. While Tifft's redshift bands do not seem to be due to contamination of observational data by light sources on the Earth (Tifft, 1978), their origin is still an open question if they are real. However, the prevalent attitude among astronomers has been to regard periodicities in z as spurious. This attitude seems basically to be justified, as can be seen by a consideration of a related class of data which has been discussed by Varshni and others.

(c) Varshni (1976) has claimed that there are significant periodicities in the numbers of objects with given redshifts z, and that these periodicities are inconsistent with a model of the Universe in which the Earth has an arbitrary location. The argument of Varshni is that either QSOs are distributed non-randomly with respect to the Earth, or else QSO redshifts are non-cosmological. Varshni had proposed a non-cosmological mechanism for QSO redshifts earlier (Varshni, 1975), suggesting that QSO emission lines might be due to laser action in the expanding envelopes of nearby stars; but this hypothesis has not been accepted. In addition, much criticism has been directed at Varshni's claim that QSO redshifts are inconsistent with a model Universe in which the Earth is randomly located. To be specific, Varshni (1976) has identified 57 groups each containing several QSOs with approximately the same z-values. The data involved in identifying these groups had been discussed by numerous people previously (see Field *et al.*, 1973), many of whom had speculated on why there appear to be exceptionally large numbers of QSOs having $z \simeq 1.95$ and other apparently preferred values of z. Varshni's argument is that, since sources with the same z-values are at the same distance (though lying in different directions) from us on the basis of Robertson/Walker cosmology, the 57 groups define 57 spherical shells with the Earth at the centre. This conclusion is based on a statistical analysis of how unlikely it would be for the QSO z-values to be distributed in the observed manner purely as a chance phenomenon. Varshni's statistical argument was criticized as incorrect by Stephenson (1977), who found that the distribution of QSO z-values is in agreement with random expectation. Varshni (1977) disagreed with the way in which Stephenson carried out the test of significance involved for the grouping of the z-values. But while the re-analysis by Varshni (1977) substantiated his previous conclusion (Varshni, 1976), it is clear that the significance of the result is in serious doubt. Two other independent analyses of the data by Owen (1977) and Weymann et al. (1978) showed the clumping of QSO redshifts to be consistent with chance fluctuations of a random distribution. This means that QSOs as a class probably do not show anomalous redshift clumping. Neither is there anything special about the values of z (e.g., $z = 1.95$) which the (probably random) clumping seems to pick out as preferred. There is no significant periodicity in QSO emission line redshifts, and no significant anisotropy either in their redshift distribution with respect to the two Galactic hemispheres (Wills and Ricklefs, 1976; see Chapter 6 also). A non-random trend in the gaps in the number/redshift distribution of QSOs was noted by Basu (1978a), but neither were the gaps periodic. Selection effects are in any case important sources of departures from randomness in the number/redshift distribution of QSOs (Basu, 1978b), despite claims (e.g., Karlsson, 1977) that periodicities exist which cannot be explained as being due to observational effects. Thus, although the nature of QSOs is still undecided (see Ginzburg and Ozernoy (1977) for a review), if they do have any anomalies at all as regards their individual redshifts then these must be of internal origin, and not to do with the locations of the sources in space or a non-Einsteinian cosmology.

The arguments of the previous three paragraphs show that, basically speaking, QSOs and galaxies have redshifts determined primarily (and probably entirely) by the Doppler effect operating in an expanding Universe. The expansion of the Universe is an effect described in the Robertson/Walker models by the scale factor S (Hubble's parameter is $H = \dot{S}/S$). However, one can if one wishes choose coordinates such that the same redshift law ($z \simeq \dot{r}/c \simeq Hr/c$) appears as an effect of the curvature of space-time, with an alternative form of the metric in which the expansion is no longer apparent (Wesson, 1978a). This does not change the solution, but only the way in which it is expressed. It does not change the fact that the redshift is understandable as an effect of Einstein's theory, and that observations of astronomical objects uphold the frame-independence of the velocity of light c

as built into that theory. (In the previously much-discussed Ritz theory, the velocity of light propagation as measured by an observer is $c + v$ where v is the velocity of the source; but astronomical tests of this dependency by Heckmann (1959, 1960), Dickens and Malin (1965) and Brecher (1977) confirmed that the velocity of light is independent of the velocity of the source, as in Einstein's theory.) As far as redshifts are concerned, there is clearly no reason for disputing that astronomical objects have z-values that are, more or less, in agreement with those predicted by simple solutions of Einstein's equations.

11.4. Non-Friedmannian Redshifts

There is only one aspect in which the Universe might reasonably be expected not to agree with the Friedmann solutions of general relativity, in which z-values are regularly distributed around an arbitrarily-located observer. This possible disagreement concerns departures from a homogeneous and isotropic distribution of matter, connected with the clumping of galaxies into clusters. (See MacCallum, 1973, and Shepley, 1973, for reviews of some anisotropic models of the Universe based on general relativity; see also Datt, 1938; Wyman, 1946, 1976, 1978; Hawkins, 1960; McVittie, 1967; McVittie and Stabell, 1967, 1968; Cahill and Taub, 1971; Bonner, 1972, 1974; Cook, 1975; Eisenstaedt, 1975a, b; Szekeres, 1975; Bonner and Tomimura, 1976; Chakravarty et al., 1976; Tomimura, 1977; Silk, 1977e; Hara, 1977; Goldman, 1978; Henriksen and Wesson, 1978a, b; and Wesson, 1978a, c, d, e, h; 1979a, b, for non-Friedmannian solutions to Einstein's equations; and see Zel'dovich, 1964; Gunn, 1967; Petrosian and Salpeter, 1968; Refsdal, 1970; Dyer and Roeder, 1972, 1973, 1974; Silk, 1974a, b; Roeder, 1975; Weinberg, 1976b; and Wesson, 1979a, for observational relations in clumpy Universe models.) A case for considering that most if not all galaxies are clumped into groups, clusters and superclusters was made by De Vaucouleurs (1958, 1970, 1971, 1975a, b, c; 1976; see also De Vaucouleurs and De Vaucouleurs, 1964; De Vaucouleurs and Corwin, 1975; De Vaucouleurs et al., 1976). This hierarchical picture was essentially confirmed by a series of 10 papers by Peebles and coworkers on the statistical analysis of catalogues of extragalactic objects using the covariance function (Peebles, 1973; Hauser and Peebles, 1973; Peebles and Hauser, 1974; Peebles, 1974a; Peebles and Groth, 1975; Peebles, 1975; Groth and Peebles, 1977; Seldner and Peebles, 1977; Fry and Peebles, 1978; Seldner and Peebles, 1978; see also Peebles 1974b, c, d; Davis, 1976; Groth et al., 1977). It seems likely that clumpiness occurs on all scales, without the existence of a large spatially homogeneous population of field galaxies (Fall et al., 1976; Soneira and Peebles, 1977; Chincarini, 1978a), as had at one time been claimed (Turner and Gott, 1975). Many galaxies are not only members of clusters but also members of smaller groups (De Vaucouleurs, 1975d; Turner and Gott, 1976a, b; Gott and Turner, 1977a, b), including binaries (Turner, 1976a, b). This makes it difficult to calculate the mean density of matter in galaxies, but it is almost certainly $\rho_U \lesssim 10^{-30}$ gm cm^{-3} (Gott and Turner, 1976; Chincarini, 1978b). It may even be $\rho_U \simeq 1 \times 10^{-32}$ gm cm^{-3} when considered as a mean value for the galaxies seen in the direction of a rich cluster like Coma, where there is a strong localized concentration of objects in what is otherwise a sparsely-populated background (Tifft and Gregory, 1976). The percentage of all galaxies that are single is almost certainly less than 18 percent (Soneira and Peebles, 1977), probably less than 10 percent (Gregory and Thompson, 1978) and could well be almost zero (Tifft and Gregory, 1976; Gregory and Thompson, 1978). Although there are slight kinks in the covariance function for some classes of data (Wesson, 1976; Tully and Fisher, 1978), it is generally smooth as far as it relates to clustering of galaxies on scales up to about 10 Mpc. This shows that galaxies, groups and clusters do not form a stepped system with well-defined scales, even though the distribution is roughly hierarchical. The break in the covariance function at a scale of about $9\ h^{-1}$ Mpc (h is H in units of 100 km sec^{-1} Mpc^{-1}) may be a result of the way in which galaxies formed (Groth and Peebles, 1977; Davis et al., 1977; Bonometto and Lucchin, 1978). Certainly, clustering continues beyond this scale, extending up to dimensions of at least $35\ h^{-1}$ Mpc (and probably beyond) as shown by computer simulations of a clumpy Universe by Soneira and Peebles (1978). Superclusters with diameters of 30–60 Mpc are known to exist (Abell, 1975; Shane, 1975; Chincarini and Rood, 1975; see also Rood and Sastry, 1971), and it would seem that a first approximation to the clustering phenomenon is a model in which there is continuous (i.e., no-scale) clumping up to dimensions of about 50 Mpc (Kiang, 1967; Peebles, 1974e; Bhavsar, 1978). Beyond that level, it is anyone's guess how the galaxies are distributed. The majority opinion is that the clustering dies out and that the Universe is homogeneous and isotropic on the very large scale. But as pointed out by Shane (1975), the postulate of isotropy has never been properly tested on the really large scale. In fact, the isotropy postulate is decidedly questionable. The distribution of superclusters is itself irregular, indicating that there could be inhomogeneities on scales of $600\ h^{-1}$ Mpc with rms density contrasts of order 0.3 (Hauser and Peebles, 1973). Further, both Shane and Peebles (see Peebles, 1978b) have noticed that in the Lick sample there is a definite large-scale variation in density over an angular scale of 40°.

The Lick catalogue of galaxies, which has been re-analysed by Seldner *et al.* (1977), extends down to an apparent magnitude of $m \simeq 19$. Therefore, if the noted variation is not due to variable obscuration in the Galaxy, it represents an inhomogeneity of almost global order.

Despite the preceding comments, many astronomers are content with the postulate that the Universe is isotropic overall. The two main reasons for this have hitherto been the observed isotropy of the 3 K microwave background (see below) and the apparent isotropy of the radio sources (as opposed to optical sources). The radio sources are, to a first approximation, distributed uniformly (see the book edited by Jauncey (1977), and Longair (1978), for reviews). Some claims of the existence of anisotropies have been made, notably by Yahil (1972), Plagemann (1973), and Maslowski *et al.* (1973) and Machalski *et al.* (1974). The latter claim has been criticized by Blake (1976), and it is in general difficult to say if any anisotropies exist in the distribution of radio sources, since the number differences involved are of the same order as uncertainties due to instrumental and observational limitations to the accuracy of the catalogues. This may explain the conflict between the results of Seldner (1977) and Seldner and Peebles (1978), who have found weak but significant clustering in 4C radio sources, and Birkinshaw (1978), who has found that radio sources lying in the directions of rich clusters of galaxies are not significantly concentrated towards the clusters. The conclusion of Seldner and Peebles (1978) has also been criticized by Masson (1978), who is of the opinion that the apparent clustering in 4C radio objects can be attributed to the absence of sources in regions obscured by the sidelobes of more intense sources and the concentration of supernova remnants near the Galactic plane. But while both of these influences are plausible ones, the matter is still controversial and all one can say with any degree of certainty is that the radio sources as seen in a projected distribution appear to be considerably less clustered than the optical sources.

One explanation for this might be that radio sources (for which redshifts are often unavailable and for which distances are therefore unknown) are intrinsically different from optical galaxies, in that the former do not cluster while the latter do. This is not a very attractive hypothesis from a morphological point of view, since there does not appear to be any sharp dividing line as regards some of their properties between ordinary galaxies, Seyfert galaxies, N-type galaxies, radio galaxies and QSOs, which might all be different phases in the life history of one class of object (Rowan-Robinson, 1977a). On the other hand, it could be that in observing many radio sources we are looking at remote objects seen as they were in the early Universe before galaxies began to clump. This brings in the questions of what the

Universe was like at early epochs and how the galaxies and clusters of galaxies formed (see the next paragraph). As far as observations of optical galaxies are concerned, clustering is undoubtedly present out to redshifts of $z \simeq 0.5$, but it is unclear if remote galaxies with redshifts of about this size are more clustered (Dodd *et al.*, 1975) or less clustered (Phillipps *et al.*, 1978) than nearer galaxies with lower redshifts. All that can be said with reasonable certainty is that the scale of clustering at $z \simeq 0.5$ is of order 1 Mpc (Dodd *et al.*, 1976), which is similar to the clustering scale of nearby rich clusters of galaxies. The extent to which models of the Universe with extended clustering agree with observation has been examined by Rainey (1977), who has evaluated number/magnitude relations for clumpy galaxy distributions; and by Wesson (1979a, b), who has investigated the dynamical parameters (H, q), the magnitude/redshift relation, and the number/redshift relation for such cosmologies. These last-mentioned relations can be made to agree with observations of optical galaxies, which as previously noted are clustered on scales of the order of the size of the Local Supercluster (30–50 Mpc) and perhaps on the scale ($\simeq 150$ Mpc) of a local third-order cluster (Dodd *et al.*, 1975; see below). However, a more interesting limit to the inhomogeneity of indefinitely-clumped cosmological models is set by radio astronomy. The most pertinent result is that of Fanti *et al.* (1978), who have investigated the isotropy of the radio sources in the B2 catalogue. The distribution of sources in this catalogue has been found to be random down to the 1σ level (i.e., no clustering of the sources) on angular scales $\gtrsim 5°$. From this, Fanti *et al.* have calculated that the three-dimensional density contrast of clustering on a characteristic scale of (say) 60 Mpc is such that $\Delta\rho/\rho_U \lesssim 10$ where ρ_U is the mean density. This result and others like it (Wesson, 1978a, 1979a, b) are of importance because they set limits on possible departures of the real distribution of astronomical sources from the uniformity that is assumed in the conventional Friedmann models of the Universe.

Constraints set by the apparent isotropy of certain classes of astronomical objects (like the radio sources), the entropy of the 3 K microwave background, the abundances of the elements, and the existence of galaxies, point to the likelihood that the Universe began as a Robertson/Walker type of singularity, with only small fluctuations of the curvature in a quiescent spacetime (Barrow and Matzner, 1977; Barrow, 1978b). This is in agreement with the model of conventional cosmology, and in disagreement with those models in which the early Universe was chaotic and clumpy on a range of scales (Misner, 1968; Barrow, 1977a; Tomita, 1977), although calculations on the production of the elements in such cosmologies do not completely rule them out (Tomita, 1972, 1973; Epstein and Petrosian, 1975; Barrow, 1976, 1977b;

Austin and King, 1977; Hartquist and Cameron, 1977; Olson, 1978; Olson and Silk, 1978). However, on the basis of the conventional big-bang model of cosmology (Harrison, 1973a), one can say that there was probably opportunity for the development of extensive clumpy structure over the history of the Universe. Reviews of the formation of galaxies and clusters of galaxies have been given by Field (1975), Zel'dovich and Novikov (1975), Jones (1976), Larson (1976), Gribbin (1976), Gott (1977) and Rees (1978c), and the subject has been extensively discussed in the books edited by Shakeshaft (1974) and Longair and Einasto (1978). There is not much consensus of opinion on this topic. Different workers attribute the main factor in the formation of clumpy structure to: gravitational interactions (Bonner, 1956a, b, 1957; Peebles, 1967, 1970, 1974b, c, d; Peebles and Yu, 1970; Gunn and Gott, 1972; Press and Schechter, 1974; Adams and Canuto, 1975; Gott and Rees, 1975; Layzer, 1975; Doroshkevich and Zel'dovich, 1975; Liang, 1976, 1977; Edwards and Heath, 1976; Fall, 1976; Fall and Saslaw, 1976; Heath, 1977; Groth *et al.*, 1977; Davis and Peebles, 1977; Davis *et al.*, 1977; Cavaliere *et al.*, 1977a, 1978; McClelland and Silk, 1977a, b, 1978; Norman and Silk, 1978; Silk and White, 1978; Perrenod, 1978; Bonometto and Lucchin, 1978; Aarseth, 1978; Fall, 1978); gas dynamical processes (Silk, 1968, 1977a, b, c; Brosche, 1970; Icke, 1973; Clutton-Brock, 1976; Kellogg, 1977; Rees and Ostriker, 1977; Cavaliere *et al.*, 1977b; Binney, 1977a, b; Binney and Silk, 1978; Rees, 1978b; White and Rees, 1978); gravitational interactions assisted by a Λ-force (Lemaître, 1961; Eisenstaedt, 1977; see also McVittie, 1933; Bonner, 1956a, 1957; Eisenstaedt, 1957a, b) or cosmological rotation (Silk, 1970; Novello and Rebouças, 1978); turbulence (Harrison, 1971; Ozernoi, 1972; Silk and Ames, 1972; Silk, 1973; Silk and Lea, 1973; Jones, 1973, 1977; Eichler, 1977); growth around primeval nuclei (Mészáros, 1974, 1975; Gribbin, 1976; Canuto, 1976, 1978; Carr, 1977a, b; Barrow and Carr, 1978); magnetic fields (Piddington, 1974; Wesson and Lermann, 1977b; Henriksen and Reinhardt, 1977; Wasserman, 1978); or phase transitions (Zel'dovich, 1963; Layzer, 1971; Hively, 1971; Layzer and Hively, 1973; Canuto, 1978). In connection with the discussion of variable G in preceding chapters, it can be pointed out here that one of the more successful accounts of cluster formation using the gravitational instability picture (Lewis, 1977) employs a time-variable value of G: with a larger value of G than the familiar one, the formation time of clusters of galaxies is shorter, and cluster ages and sizes lead one to believe that G may depend on time as $G \propto t^{-1/2}$ on the basis of this model. Although there is little agreement about how the real galaxies and clusters of galaxies might have formed, it has been shown by Peebles (1978c) that a hierarchical clustering pattern, once set up, would have been stable against disruption over cosmological time scales, contrary to an earlier claim by Rees (1977b). In the present Universe, clumpiness is a common phenomenon that must, at some level, cause small departures from the large-scale, regular distribution of redshifts due to the Hubble expansion. The question being treated in this chapter is whether there exist any non-Hubble redshifts that cannot be satisfactorily accounted for in terms of the clumping of galaxies and other already understood astrophysical phenomena.

A systematic directional discrepancy between the redshifts of a class of ScI-type galaxies, with velocities in the range 4000–7500 km sec^{-1}, and the redshift dependency of the Hubble law was discovered by Rubin, Ford and Rubin (1973) and has come to be known as the Rubin/Ford/Rubin anomaly. This RFR anomaly might, it is true, be due to a departure from Einstein's theory, but it might also be due to other things (see below). It was suggested by Rubin, Ford and Rubin (1973) that the redshift anomaly owes its existence to the fact that Hubble's parameter H is slightly different in two different directions in space. This may well be the case, and whatever its origin the nature of the RFR anomaly evidently represents a major question in cosmology. Its reality has been confirmed by Jaakkola *et al.* (1975) using a sample of compact galaxies with absorption spectra, and by Le Denmat *et al.* (1975) using a class of Sc-type galaxies calibrated with van den Bergh's absolute magnitudes. Guthrie (1976) has likewise confirmed its existence by using the brightest galaxies in clusters, and has shown that it extends out to sources with velocities of up to 25 000 km sec^{-1} (i.e., to scales of 250–500 Mpc depending on the value of H). While it has been claimed that the RFR anomaly could be due to bias of one form or another in the data, this claim has not been validated (see below), and Karoji (1978) has shown that it must be due to a process affecting the apparent magnitudes, absolute magnitudes or velocities of recession of the galaxies involved. That is, the RFR redshift anomaly appears to represent a physical property of the galaxies concerned rather than something to do with the way in which they are sampled.

Some time after the first study of the anisotropic distribution of the redshifts of ScI galaxies by Rubin *et al.* (1973), similar data were used by Rubin *et al.* (1976a, b) to find the peculiar velocity of the Milky Way with respect to other, more distant galaxies that are assumed to represent a cosmological reference field (the comoving frame). This way of expressing the data is equivalent to that using an anisotropy in H, except that the emphasis is now put on the velocity of the Galaxy and the Local Group rather than on the remoter ScI galaxies employed in the study. After correcting for other effects (e.g., the motion of the Sun in the Galaxy), Rubin *et al.* (1976b) concluded that the Galaxy and the Local Group are moving at 454 (± 125) km sec^{-1} toward the direction given by Galactic coordinates $l = 163°, b = -11°$. As noted above, this is a relative velocity between us and distant galaxies, and could be due to various causes (to be discussed below), including an anisotropy in H. But irrespective of its precise origin, if this is a real velocity effect (and not a spurious effect caused by absorption or variations in galaxy magnitudes), then it brings in a conflict with the apparent isotropy of the 3 K background. Rubin *et al.* (1976b) concluded that the effect is due to a velocity of some kind, and so one must examine how this velocity is to be reconciled with the isotropy of the 3 K microwave background.

As mentioned in Chapters 2 and 3, the microwave background is believed to be the remnant of the big-bang origin of the Universe. The field is observed to be isotropic over a range of angular scales down to a level of order $\Delta T/T \approx 10^{-3}$, as reviewed in the books edited by Longair (1974) and Longair and Einasto (1978) and in Wesson (1979a, b). The shortcomings of attempts at accounting for the origin and isotropy of the background on non-cosmological bases have been discussed in Chapter 3 and in Peebles (1971a), Wesson (1975b, 1978a) and Wesson and Lermann (1976). The hypothesis of an all-pervasive cosmological nature for the field has so far survived the numerous attacks which have been made on it. The only outstanding problem is that of the expected cutoff, at an energy of about 6×10^{19} eV, in the spectrum of high-energy cosmic rays due to scattering of the latter off the photons of the (assumed universal) 3 K field (Wdowczyk and Wolfendale, 1975; Wesson, 1978a; Chapter 10). Some cosmic rays with energies in the range $1-3 \times 10^{20}$ eV are observed, so if the cutoff does not manifest itself when more data on high-energy particles become available, the assumption of an all-pervasive intergalactic microwave photon field might be compromised. But this eventuality can also be avoided by assuming that the cosmic rays do not travel cosmological distances through the 3 K field, but instead have a local origin, in the halo of the Galaxy or in the Local Supercluster, for example (Wdowczyk and Wolfendale, 1975). Accepting the usual, cosmological origin for the field, it is natural to suppose that its rest frame is synonymous with the comoving frame of cosmological theory. Therefore, any departure of the motion of the Galaxy from the comoving frame defined by the other galaxies ought to show up as a Doppler anisotropy in the temperature of the 3 K background of order v_{pec}/c, where v_{pec} is the peculiar velocity of the Galaxy (and Local Group) as found by Rubin *et al.* (1976a, b). The expected anisotropy would be of order $\Delta T/T \approx 454/(3 \times 10^5) \approx 1.5 \times 10^{-3}$, with its exact value depending on how one corrects for other velocities with respect to the comoving frame and on the directions involved. Further, the anisotropy should be a dipole one if it is due to a velocity effect, rather than a localized anisotropy of the type which one might expect to be associated with other, non-cosmological effects. The latter include effects to do with the scattering of microwave photons off hot electrons in clusters of galaxies (Sunyaev and Zel'dovich, 1972; Gull and Northover, 1976; Lake and Partridge, 1977; Birkinshaw et al., 1978a, b; Fabbri et al., 1978), dust in the Galaxy (Fazio and Stecker, 1976; Forman, 1977), long-wavelength gravitational waves (Anile and Motta, 1978), and the clumpy nature of the medium from which the radiation was last scattered (Longair and Einasto, 1978; Wesson, 1979a, b). Indeed, the background as it relates to the last of these effects is known to be isotropic over some scales even down to $\Delta T/T \approx 10^{-4}$ (Caderni et al., 1977; Boynton, 1978; Longair and Einasto, 1978; Wesson, 1979a). The question is rather if there exists an anisotropy in the 3 K microwave background that is of order $\Delta T/T \approx 10^{-3}$ and is of global (two-hemisphere) nature.

An anisotropy of this type and of the right order of magnitude was discovered by Partridge and Wilkinson (1967), Conklin (1969) and Henry (1971). The first two of these references only defined the anisotropy in one plane (i.e., they defined only one component of the Earth's velocity with respect to the radiation, assuming that the anisotropy is interpreted as a Doppler effect), while the second defined components in two senses which allowed the magnitude and direction to be uniquely determined. However, the discovery of this background anisotropy did not cause the concern which attended the later discovery of the RFR anomaly. The reason was that most of the anisotropy could be removed by making use of a reference frame defined by nearby galaxies rather than by the Earth, so it was felt that the anisotropy did not constitute a threat to the conventional isotropic cosmological models, but was instead something which had mainly to do with the motions of galaxies in the neighbourhood of the Milky Way. The argument involved for adopting such an adjustment of the anisotropy goes as follows: The anisotropy is measured from the Earth, but for cosmological purposes one wishes to know if there is an anomaly with respect to the comoving frame of the galaxies as a whole, so one adds in vectors expressing the velocity of the Earth around the Sun (small), the velocity of the Sun in the Milky Way (substantial), the velocity of the Milky Way in the Local Group of galaxies (small) and the velocity of the Local Group in the Local Supercluster (substantial). If one stops at this stage, the amount of the anisotropy which is left over is not very large. This can be interpreted as meaning that the Local Supercluster itself has a negligible peculiar velocity with respect to the more remote galaxies and the frame in which the 3 K background is isotropic. (Actually, the frame in which it is isotropic is in practice connected with the frame in which it was last scattered; but apart from small perturbations due to localized, primeval condensations, it is usually assumed that the frame of last scattering in the early Universe corresponds to the comoving frame in which the galaxies of the present Universe are at rest.) The fact that there was no significant anisotropy left when an adjustment was made to the frame of the Local Supercluster as just outlined was congenial to most cosmologists. It meant that the 3 K microwave background was isotropic with respect to a frame defined by the large-scale (i.e., larger than supercluster scale) distribution of the galaxies. (The residual anisotropy depends considerably on the somewhat uncertain dynamics of the Local Supercluster (De Vaucouleurs, 1958; De Vaucouleurs and Peters, 1968; Sciama, 1967; Stewart and Sciama, 1967; Conklin, 1969; Henry, 1971), but in velocity terms it is probably not more than 100 km sec^{-1}.) However, in view of recent data, discussed previously in this chapter, which tend to show that galaxies may be clumped on larger than supercluster scales, one must now ask if the 'adjustment' to near zero of the 3 K anisotropy as described above is not merely fortuitous. In other words, one must ask if there could be a larger cluster, of which the Local Supercluster is a part, with respect to which the anisotropy is not nearly zero but instead finite.

This is certainly possible on the basis of modern data on the distribution of the galaxies. If it should turn out that there is a cluster in which the Local Supercluster is contained and has a notable peculiar velocity, then the situation would be that the 3 K microwave background is, after all, anisotropic. The situation is completely undecided. The question of whether the anisotropy found by Partridge and Wilkinson, Conklin, and Henry represents a disproof of the assumption that the 3 K field is isotropic with respect to the comoving frame, depends entirely on the degree to which the galaxies are clustered. If the ambiguity in defining a scale on which the galaxies are at rest with respect to the 3 K field continues to be present as more data accumulate on the large-scale distribution of the galaxies, this will mean that in practice it is impossible to define a comoving frame for the galaxies at all. That is, in the absence of any proof or guarantee that clustering dies out on large scales, there would be no sense in trying to define a comoving frame as far as the dynamics of the galaxies are concerned. (One could, by fiat, define the comoving frame to be that in which the 3 K field is isotropic; but in the presence of unlimited clustering this would be a purely academic statement of no practical significance.) Actually, the difficulty of knowing where to stop in trying to 'adjust' the anisotropy to zero by finding a frame in which the galaxies are at rest with respect to the 3 K microwave background should not cause surprise. Apart from using the 3 K background itself, it is impossible to define an absolute standard of rest in the uniform, nearly empty Robertson/Walker models of the Universe (Fox et al., 1975; Von Hoerner, 1973b), although it is possible to measure parallax caused by the motions of the galaxies relative to each other (Noerdlinger, 1977). Geodesic properties of the space-time background of the Universe do not in practice provide a frame which one can use as a reference with which to compare the RFR anomaly or anisotropies in the 3 K background. The reason is that local effects of the clumpy distribution of the galaxies swamp the effects of the metric as this is related to the mean gravitational field of all the matter in the Universe. From the point of view of wishing to compare the RFR galaxy redshift anomaly with possible anisotropies in the 3 K microwave background, it is seen to be better to make a direct comparison, rather than trying to use as intermediary standard a comoving rest frame which is of little practical value and might in principle be impossible to define except in the most abstract way.

Smoot et al. (1977) confirmed that there is a dipole anisotropy in the 3 K background, and the relation of this to the RFR anomaly has been discussed by Rowan-Robinson (1977b). The anisotropy has been discussed in the context of a velocity of the Galaxy and the Local Group, so it can be compared to the RFR anomaly (see below). As seen directly from the Earth, the maximum of the anisotropy is of size $3.5 (\pm 0.6) \times 10^{-3}$ K in the direction towards 11 (± 0.5) hr R.A., 6 (± 10)° dec., and shows that the Earth is moving with respect to the 3 K radiation. The effect is of the same order of magnitude and is in roughly the same direction as found by Partridge and Wilkinson (1967), Conklin (1969) and Henry (1971). It also agrees with the slightly less accurate results of Corey and Wilkinson (1976). When one adjusts for the velocity of the Earth around the Sun and the Sun around the centre of the Galaxy, one can interpret the temperature anisotropy in terms of a peculiar velocity of the Galaxy and the Local Group with a magnitude of 600 km sec^{-1} in the direction $l = 261°$, $b = +33°$. The size of this anisotropy velocity is comparable to the size of the redshift-anomaly velocity of the RFR effect (450 km sec^{-1}), but the two directions involved are roughly at right angles to each other. There is little doubt that the microwave anisotropy is of cosmological significance since it is quite accurately dipole in nature (Muller, 1978). The question therefore arises of how the peculiar velocity of the Galaxy with respect to the 3 K microwave background frame comes to be at right angles to the peculiar velocity of the Galaxy with respect to the redshift frame of the remote galaxies.

Muller (1978) suggested that the discrepancy can be understood as a combination of a peculiar velocity of the Galaxy of 600 km sec^{-1} with respect to the microwave background, plus a peculiar velocity of the whole RFR class of galaxies of 800 km sec^{-1} also with respect to the microwave background, the two velocities being offset by 33°. (I.e., there is a right-angled triangle of velocities with respect to the 3 K background, this triangle having sides of lengths 600 km sec^{-1} and 450 km sec^{-1} which enclose a right angle, while the hypotenuse is of length 800 km sec^{-1}.) In this way, the RFR velocity of 450 km sec^{-1} can be seen as the resultant of two peculiar velocities of 600 km sec^{-1} and 800 km sec^{-1}, both with respect to the microwave background. This hypothesis accounts for the dynamical facts; but involves the movement of a segment of space-time of large size (the RFR sample has a depth of about 100 Mpc) at a velocity which is considerably in excess of the value $v_{pec} \lesssim 50$ km sec^{-1} usually regarded as a limit for a typical galactic peculiar velocity. (There is some controversy about galactic peculiar velocities, but the general opinion is that they do not exceed the quoted value. For example, Sandage and Tamman (1974) could not detect any peculiar velocity of the Virgo cluster, and concluded (Sandage and Tamman, 1975a) that the local velocity field of the galaxies is a quiet Hubble flow in which peculiar velocities are $v_{pec} \lesssim 50$ km sec^{-1}. Rowan-Robinson (1977b) also took peculiar velocities to be no larger than this, and so was forced to suggest that the Smoot *et al.* (1977) result indicated that entire superclusters might be in motion with respect to each other. Recently, Gudehus (1978) arrived at a peculiar velocity of 658 ($\pm$96) km sec^{-1} for the Virgo cluster inside the Local Supercluster, but while this might represent a peculiar velocity with respect to distant galaxies if the Local Supercluster is unbound, it might also represent a dispersion velocity of a cluster within a bound supercluster.) In view of the very large peculiar velocities which it implies, the hypothesis of Muller (1978) must be considered doubtful when judged on the basis of presently available data on the velocity dispersion of galaxies with respect to the Hubble flow.

While there is doubt about the reason for the sizes of the velocity terms involved, the reason for the *offset* in velocity directions probably involves some feature of the large-scale structure of the Universe. It cannot, for example, be easily accounted for by motions of expansion and rotation inside the Local Supercluster, at rates of a few 100 km sec^{-1} (De Vaucouleurs, 1958; Rowan-Robertson, 1977b), in view of the sizes and directions of the vectors involved. (In theory, the microwave background frame and the redshift frame of the distant RFR galaxies ought to be the same; it is the fact that they apparently are not the same which makes it of little interest to reduce one or the other of the two vectors involved to zero by a suitable choice of Local Supercluster dynamics as was done with the early 3 K background results of Partridge and Wilkinson, Conklin, and Henry discussed above.) One possible explanation for the approximately 90° offset of the two vectors is given by a cosmological model of Wesson (1979a). This model, which will be explained in more detail below, is characterized by different dynamical properties in two perpendicular directions (the radial and azimuthal directions in terms of the metric). The model is inhomogeneous, so an observer looking along the radial direction in the senses toward and away from a centre of inhomogeneity could expect to see slightly different temperatures for the 3 K background in the two senses of view (physically, the reason is that the temperature of the 3 K background depends on the gravitational potential, i.e., on g_{ij}, which is inhomogeneous for finite epochs in this model). This effect could account for the 3 K background anisotropy of Smoot *et al.* (1977). The RFR anomaly would be accounted for by the circumstance that the inhomogeneity causes galaxies seen in the radial and azimuthal dirctions to move in slightly different ways (see below; the dispersion in velocities in either one of the two directions could still be small). Thus, an observer would see a 3 K anisotropy in the radial direction and, compared to galaxy motions in that direction, a redshift anomaly in the azimuthal direction at right angles to it. This explanation of the data, like that of Muller, must be considered speculative.

A more conservative approach to the contradiction in directions between the 3 K anisotropy and the redshift anomaly is to assume that the former can be adjusted to zero at some level (e.g., that of the Local Supercluster) while the latter has, contrarily, some non-adjustable cause. This is a reasonable attitude to adopt – if it works. Thus, the main problem facing conventional cosmological models in this area is to find a workable hypothesis for the nature of the RFR redshift anomaly.

Explanations of the RFR anomaly are numerous, and can conveniently be divided up into those involving (1) absorption; (2) the effects of galaxy clustering in an (on average) homogeneous Universe; and (3) H-anisotropy in a globally inhomogeneous Universe.

(1a) Jaakkola *et al.* (1975b) and Guthrie (1976) discussed the possibility that the anomaly might be caused by intergalactic (or possibly Galactic) absorption, but this does not seem able to account for all of the effect. (A related process (1b) in which photons from distant sources pass through and interact with the radiation fields of remote clusters was mentioned by Le Denmat *et al.* (1975) and discussed by Jaakkola *et al.* (1976), but this seems equally unlikely.) However, (1c) Teerikorpi (1978) made a case for believing that dust and gas in the Galaxy might play a part in accounting for the anomaly: unevenness in absorption, possibly connected with the Orion spiral arm, can account for a difference of up to 0.4 mag. in the brightnesses of extragalactic sources as seen in different directions, and this might reasonably produce an effect similar to the observed RFR anomaly.

(2a) Sandage and Tamman (1975a, b) discussed the RFR anomaly in terms of possible departures from uniformity of the velocity field of the galaxies due to the clumpy distribution of the latter. At first, they tried to make a case for believing that the RFR anomaly was an effect only of bias in the sample of remote ScI galaxies in which the effect was originally found, the velocity field being regular with no depatures from the Hubble law with a well-defined value of H (Tamman and Sandage, 1975a). This work was severely criticized by De Vaucouleurs (1976), who showed that, as far as local galaxies are concerned, the velocity field is certainly not uniform but is instead noticeably affected by the Local Supercluster (see also the discussion of Segal's theory in Chapter 9). The existence of the Supercluster, which has a diameter in the range 50–100 Mpc, depresses the value of H below its free-space, asymptotic value ($H_0 \simeq 75$ km sec^{-1} Mpc^{-1}), producing an anisotropy in the velocity field of nearby galaxies and in the local m/z relation. In principle, a similar effect could account for the RFR anomaly if there are many clumps like the Local Supercluster; and Sandage and Tamman (1975b) did admit that, while they still believed the anomaly to be caused by a sampling bias, it could also be due to cluster-produced anisotropies in the velocity field of the galaxies. Following Rubin *et al.* (1973), Sandage and Tamman (1975b) noted that the data can be expressed as an anisotropy in H, with $H_{II}/H_I \simeq 1.25$ for the two different directions involved. This departure from the uniform velocity field of conventional cosmological models (in which H is the same in all directions) could be explained by the existence of localized clumps of galaxies on a scale of 100 Mpc situated in a background that would be uniform on larger scales. The clumps would cause perturbations in the otherwise smooth gravitational potential in which the galaxies move, so accounting for the departure of their velocities from a uniform flow. (2b) Doroshkevich and Shandarin (1976) suggested that the anisotropy in H on scales of 100–300 Mpc could be a dynamical effect due to the anisotropic motions of galaxies within a large pancake-shaped cluster that formed as a result of the growth of non-linear gravitational instabilities in the early Universe. (2c) A related suggestion by Fennelly (1976) is that the anisotropy in H could be due to the existence of an inhomogeneity in the Universe which resembles a clump of diameter 540 Mpc and in which the Galaxy is located 120 Mpc from the centre (this lying in the direction towards R.A. 12 hr, dec. 20°). Fennelly used a self-similar model for the clump to estimate that its central regions might be observable as a zone in which the 3 K microwave background would be expected to show temperature perturbations $\Delta T/T$ or order 10^{-3}–10^{-4}.

(2d) Fall and Jones (1976) disagreed with the basis of the three previous accounts, namely that the RFR anomaly involves a departure of the velocities of the galaxies concerned from a uniform flow. Fall and Jones instead attempted to explain the anomaly as being an effect purely of the inhomogeneous distribution of the galaxies in the sample of Rubin *et al.* (1973). It was suggested by Fall and Jones that these galaxies are clumped into two main clouds but are moving as members of a uniform flow in a low-density Universe in which H is not noticeably direction-dependent. (2e) A possible explanation of the RFR anomaly that has something in common with both (2a)–(2c) and (2d) is suggested by a discussion of a similar problem by Fairall (1978). In the Virgo cluster, there is a slight difference between the redshifts of the spiral galaxies and the redshifts of the elliptical and S0 (lenticular) galaxies. This does not necessarily mean that there is something intrinsically different about the redshift-producing mechanism as it involves these two classes of objects. The difference can be explained as a dynamical one caused by the fact that the two classes of galaxies are concentrated around slightly different centres in the cluster. Thus, there is no longer any reason for believing that the observed slight redshift differences between galaxies of different types (Jaakkola, 1971; De Vaucouleurs, 1972b, 1961; De Vaucouleurs and De Vaucouleurs, 1963) – from ellipticals through lenticulars to spirals – have anything to do with redshift anomalies that are properties of the galaxies themselves. On the contrary, they are probably just the product of clustering that is asymmetric in the dynamical sense. (A contribution to a redshift anomaly produced in this way might also come from redshifts caused by velocities transverse to the line of sight to the centre of mass, as suggested by Harrison (1975) for the case of groups of galaxies.) The hypothesis of Fairall, besides providing an explanation for the redshift discrepancy in the Virgo cluster, also helps one to understand the similar problems which are present in groups of galaxies (Jaakkola and Moles, 1976) and the Local Supercluster (Teerikorpi and Jaakkola, 1977). It is plausible that, applied on a larger scale, it might in addition account for the RFR anomaly if galaxies are clumped into clusters of very large size.

(3) The explanations in classes (1) and (2) are all consistent with the belief that the real Universe is described by a homogeneous Robertson/Walker model based on Einstein's theory of general relativity. (This is true even if the model has clumps, provided that the clumpiness dies out on sufficiently large scales.) By retaining general relativity but dropping the homogeneity

assumption, it is possible to find solutions to Einstein's equations that represent indefinitely-clustered models of the Universe (Wesson, 1978a, d; 1979a, b). These solutions are the analogues in general relativity of the hierarchical models of Newtonian theory, in which clusters are contained in larger clusters, and so on without limit (De Vaucouleurs, 1970). One such relativistic hierarchical model, already mentioned above, has been worked out in detail (Wesson, 1979a). It has the property that there are two different values of H in directions at right angles to each other for all epochs of the Universe t in which t is finite ($t \not\to \infty$). By a suitable choice of parameters, one can obtain agreement with the data of Rubin, Ford and Rubin.

The three main ways of accounting for the RFR redshift anomaly outlined in the preceding three classes are all, to varying degrees, plausible. Those falling into class (2), and in particular (2a)–(2c), represent very realistic ways of explaining the observed departure from Hubble's law in its isotropic form. There is definitely no reason for believing that the RFR anomaly indicates a breakdown of Einsteinian cosmology. It is probably just a result of the large-scale clustering of ordinary galaxies.

Further, the RFR anomaly is the only redshift anomaly that has been substantiated in astrophysics. The discussion of the first part of this chapter, concerning the tired-light hypothesis for the redshift and the claims which have been made by Arp, Tifft and Varshni for the existence of anomalous redshifts, leaves one unconvinced that there is anything untoward that needs to be explained. The Universe may be non-Friedmannian insofar as the galaxies are distributed inhomogeneously; but while clustering can produce departures of galaxy redshifts from the simple Hubble law, it does not upset the belief that the redshifts themselves are due to the Doppler effect. The RFR anomaly has been confirmed as being real in the sense that it represents a departure from the Hubble law which may possibly be connected with a large peculiar velocity of the Galaxy with respect to the frame of the 3 K background radiation. But that anomaly is an anomaly in the dynamical (as opposed to the electromagnetic) sense, and probably has a dynamical explanation. This means that there is no evidence of *any* kind at the present time that leads one to doubt that the galactic redshift is understandable as the result of the Doppler effect acting in an expanding, general-relativistic Universe.

REFERENCES

Aarseth, S. J.: 1978, in *The Large Scale Structure of the Universe* (I.A.U. Symp. 79, eds. M. Longair and J. Einasto), pp. 189–196, D. Reidel, Dordrecht, Holland.
Abell, G. O.: 1975, in *Galaxies and the Universe* (Vol. 9, *Stars and Stellar Systems*, eds. A. Sandage, M. Sandage, and J. Kristian), pp. 601–645, Univ. Chicago Press, Chicago, Ill.
Abers, E. S. and Lee, B. W.: 1973, *Phys. Rep.* **9**, 1–141.
Adams, P. J.: 1978, *J. Gen. Rel. Grav.* **9**, 53–57.
Adams, P. J. and Canuto, V.: 1975, *Phys. Rev.* **D12**, 3793–3799.
Aldrovandi, R., Caser, S., and Omnés, R.: 1973, *Nature* **241**, 340–341.
Alfvén, H. and Mendis, A.: 1977, *Nature* **266**, 698–699.
Altschul, M. S.: 1978, *Found. Phys.* **8**, 69–92.
Alway, G.: 1969, *Nature* **224**, 155–156.
Ambartsumian, V. A.: 1961, *Astron. J.* **66**, 536–540.
Anderson, J. D., Keesey, M. S. W., Lau, E. L., Standish, E. M., and Newhall, X.: 1978, *Acta Astronautica* **5**, 43–61.
Anderson, J. L.: 1971, *Phys. Rev. (Ser. 3)* **D3**, 1689–1691.
Angeletti, L. and Giannone, P.: 1978, *Astron. Astrophys.* **70**, 523–529.
Anile, A. M. and Motta, S.: 1978, *Monthly Notices Roy. Astron. Soc.* **184**, 319–326.
Antonav, V. A. and Chernin, A. D.: 1977, *Astrophysics* **13**, 145–146.
Arcidiacono, G.: 1976, *J. Gen. Rel. Grav.* **7**, 885–889.
Arnowitt, R., Deser, S., and Misner, C. W.: 1960, *Phys. Rev. (Ser. 2)* **120**, 313–320.
Arp, H.: 1966, *Atlas of Peculiar Galaxies*, California Inst. of Technology, Pasadena, Calif.
Arp, H.: 1970, *Nature* **225**, 1033–1035.
Arp, H.: 1971a, *Science* **174**, 1189–1200.
Arp, H.: 1971b, *Nature Phys. Sci.* **231**, 103–104.
Arp, H.: 1974a, in *Confrontation of Cosmological Theories with Observational Data* (I.A.U. Symp. 63, ed. M. Longair), pp. 61–67, D. Reidel, Dordrecht, Holland.
Arp, H.: 1974b, in *The Formation and Dynamics of Galaxies* (I.A.U. Symp. 58, ed. J. R. Shakeshaft), pp. 199–224, D. Reidel, Dordrecht, Holland.
Arp, H. and Madore, B. F.: 1975, *Observatory* **95**, 212–214.
Aspden, H.: 1966, *The Theory of Gravitation* (2nd. ed.), Sabberton Publications, Southampton, U.K.
Aspden, H. and Eagles, D. M.: 1972, *Phys. Lett.* **A41**, 423–424.
Atkin, R. H. and Bastin, E. W.: 1970, *Int. J. Theoret. Phys.* **3**, 449–466.
Atkinson, R. d'E.: 1963, *Proc. Roy. Soc. Lond.* **A272**, 60–78.
Atkinson, R. d'E.: 1965a, *Astron. J.* **70**, 513–516.
Atkinson, R. d'E.: 1965b, *Astron. J.* **70**, 517–523.
Austin, S. M. and King, C. H.: 1977, *Nature* **269**, 782.
Babala, D.: 1975, *J. Phys.* **A8**, 1409–1412.
Bailey, V. A.: 1959, *Nature* **184**, 537.
Bailey, V. A.: 1960a, *Nature* **186**, 508–510.
Bailey, V. A.: 1960b, *J. Atmos. Terr. Phys.* **18**, 256.
Bailey, V. A.: 1961a, *Nature* **189**, 43–44.
Bailey, V. A.: 1961b, *Nature* **189**, 44–45.

Bailey, V. A.: 1962, *Nature* **194**, 649.
Bailey, J., Borer, K., Combley, F., Drumm, H., Krienen, F., Lange, F., Picasso, E., von Ruden, W., Farley, F. J. M., Field, J. H., Flegel, W., and Hattersley, P. M.: 1977, *Nature* **268**, 301–305.
Bakesigaki, A. and Inomata, A.: 1971, *Nuovo Cimento (Lett., Ser. 2)* **2**, 697–700.
Ball, J. A., Dickinson, D. F., Lilley, A. E., Penfield, H, and Shapiro, I.: 1970, *Science* **167**, 1755–1757.
Bally, J. and Harrison, E. R.: 1978, *Ap. J.* **220**, 743–744.
Barker, B. M.: 1978, *Astrophys. J.* **219**, 5–11.
Barker, K. D.: 1977, *Astron. Astrophys.* **60**, 291.
Barnes, R. C. and Prondzinski, R.: 1972, *Astrophys. Sp. Sci.* **18**, 34–39.
Barnothy, J. M. and Tinsley, B.: 1973, *Astrophys. J.* **182**, 343–349.
Barrow, J. D.: 1976, *Monthly Notices Roy. Astron. Soc.* **175**, 359–370.
Barrow, J. D.: 1977a, *Nature* **267**, 117–120.
Barrow, J. D.: 1977b, *Monthly Notices Roy. Astron. Soc.* **178**, 625–649.
Barrow, J. D.: 1978a, *Monthly Notices Roy. Astron. Soc.* **184**, 677–682.
Barrow, J. D.: 1978b, *Nature* **272**, 211–215.
Barrow, J. D. and Carr, B. J.: 1978, *Monthly Notices Roy. Astron. Soc.* **182**, 537–558.
Barrow, J. D. and Matzner, R. A.: 1977, *Monthly Notices Roy. Astron. Soc.* **181**, 719–727.
Barut, A. O. and Böhm, A.: 1965, *Phys. Rev. (Ser. 2)* **B139**, 1107–1112.
Bastin, E. W.: 1971, in *Quantum Theory and Beyond* (ed. E. W. Bastin), pp. 213–226, Cambridge Un. Press, London.
Basu, D.: 1978a, *Astrophys. Space Sci.* **53**, 251–256.
Basu, D.: 1978b, *Nature* **273**, 130–131.
Bateman, H.: 1909, *Proc. London Math. Soc.* **7**, 70–89.
Bateman, H.: 1910, *Proc. London Math. Soc.* **8**, 223–264.
Baum, W. A. and Florentin-Nielsen, R.: 1975, *Bull. Am. Astron. Soc.* **7**, 412.
Baum, W. A. and Florentin-Nielsen, R.: 1976, *Astrophys. J.* **209**, 319–329.
Beck, A. E.: 1961, *J. Geophys. Res.* **66**, 1485–1490.
Beck, A. E.: 1969, in *The Application of Modern Physics to the Earth and Planetary Interiors* (ed. S. K. Runcorn), pp. 77–83, J. Wiley, London.
Bég, M. A. B.: 1974, in *Fundamental Theories in Physics* (eds. S. L. Mintz, L. Mittag, and S. M. Widmayer), pp. 121–151, Plenum Press, New York.
Bekenstein, J. D.: 1977, *Phys. Rev.* **D15**, 1458–1468.
Bekov, A. A. and Omarov, T. B.: 1978, *Sov Ast., Lett.* **4**, 16–18.
Bellert, S.: 1969, *Astrophys. Space Sci.* **3**, 268–282.
Bellert, S.: 1970, *Astrophys. Space Sci.* **7**, 211–230.
Bellert, S.: 1977, *Astrophys. Space Sci.* **47**, 263–276.
Bergmann, P. G.: 1953, *Introduction to the Theory of Relativity*, Prentice-Hall, New York.
Bergmann, P. G.: 1968, *Int. J. Theor. Phys.* **1**, 25–36.
Bhamra, K. S., Finkelstein, A. M., Kreinovich, V. J., and Gurevich, L. E.: 1978, *Astrophys. Space Sci.* **57**, 371–380.
Bhavsar, S. P.: 1978, *Astrophys. J.* **222**, 412–420.
Bicknell, G. V.: 1974a, *J. Phys.* **A7**, 341–343.
Bicknell, G. V.: 1974b, *J. Phys.* **A7**, 1061–1069.
Bicknell, G. V.: 1976, *J. Phys.* **A9**, 1077–1080.
Bicknell, G. V. and Henriksen, R. N.: 1978a, *Astrophys. J.* **219**, 1043–1057.
Bicknell, G. V. and Henriksen, R. N.: 1978b, *Astrophys. J.* **225**, 237–251.
Bicknell, G. V. and Klotz, A. H.: 1976a, *J. Phys.* **A9**, 1637–1645.
Bicknell, G. V. and Klotz, A. H.: 1976b, *J. Phys.* **A9**, 1647–1654.
Binney, J.: 1977a, *Astrophys. J.* **215**, 483–491.

Binney, J.: 1977b, *Astrophys. J.* **215**, 492–496.
Binney, J. and Silk, J.: 1978, *Comm. Astrophys.* **7**, 139–149.
Birkhoff, G.: 1950, *Hydrodynamics*, Princeton Univ. Press, Princeton, N.J.
Birkinshaw, M.: 1978, *Monthly Notices Roy. Astron. Soc.* **184**, 387–396.
Birkinshaw, M., Gull, S. F., and Northover, K. J. E.: 1978a, *Nature* **275**, 40–41.
Birkinshaw, M., Gull, S. F., and Northover, K. J. E.: 1978b, *Monthly Notices Roy. Astron. Soc.* **185**, 245–262.
Birrel, N. D. and Davies, P. C. W.: 1978, *Nature* **272**, 35–37.
Bishop, N. T.: 1976, *Monthly Notices Roy. Astron. Soc.* **176**, 241–247.
Bishop, N. T. and Landsberg, P. T.: 1976, *Nature* **264**, 346–347.
Blackett, P. M. S.: 1947, *Nature* **159**, 658–666.
Blackett, P. M. S.: 1949, *Phil. Mag. (Ser. 7)* **40**, 125–150.
Blair, A. G.: 1974, in *Confrontation of Cosmological Theories with Observational Data* (I.A.U. Symp. 63, ed. M. Longair), pp. 143–155, D. Reidel, Dordrecht, Holland.
Blake, G. M.: 1976, *Monthly Notices Roy. Astron. Soc.* **174**, 63P–68P.
Blake, G. M.: 1977, *Monthly Notices Roy. Astron. Soc.* **178**, 41P–43P.
Blake, G. M.: 1978, *Monthly Notices Roy. Astron. Soc.* **185**, 399–407.
Blinnikov, S. I.: 1978, *Astrophys. Space Sci.* **59**, 13–17.
Blin-Stoyle, R. J.: 1975, *Nature* **257**, 179–180.
Bondi, H.: 1961, *Cosmology* (2nd. ed.), p. 149, Cambridge Univ. Press, London.
Bonner, W. B.: 1956a, *Z. Astrophys.* **39**, 143–159.
Bonner, W. B.: 1956b, *Monthly Notices Roy. Astron. Soc.* **116**, 351–359.
Bonner, W. B.: 1957, *Monthly Notices Roy. Astron. Soc.* **117**, 104–117.
Bonner, W. B.: 1964, *Monthly Notices Roy. Astron. Soc.* **128**, 33–47.
Bonner, W. B.: 1972, *Monthly Notices Roy. Astron. Soc.* **159**, 261–268.
Bonner, W. B.: 1974, *Monthly Notices Roy. Astron. Soc.* **167**, 55–61.
Bonner, W. B. and Tomimura, N.: 1976, *Monthly Notices Roy. Astron. Soc.* **175**, 85–93.
Bonometto, S. A. and Lucchin, F.: 1978, *Astron. Astrophys.* **67**, L7–L9.
Born, M.: 1954, *Proc. Phys. Soc.* **A67**, 193–194.
Bottinelli, L. and Gougenheim, L.: 1973, *Astron. Astrophys.* **26**, 85–89.
Bourgin, D.: 1936, *Phys. Rev. (Ser. 2)* **50**, 864–868.
Bouvier, P. and Maeder, A.: 1978, *Astrophys. Space Sci.* **54**, 497–508.
Boynton, P.: 1978, in *The Large Scale Structure of the Universe* (I.A.U. Symp. 79, eds. M. Longair and J. Einasto), pp. 317–326, D. Reidel, Dordrecht, Holland.
Braginskii, V. B., Caves, C. M., and Thorne, K. S.: 1977, *Phys. Rev. (Ser. 3)* **D5**, 2047–2068.
Braginskii, V. B. and Panov, V. I.: 1972, *Soviet Phys. JETP* **34**, 463–466.
Braginskii, V. B. and Ginzburg, V. L.: 1974, *Soviet Phys. Dokl.* **19**, 290–291.
Bramson, B. D.: 1974, *Phys. Lett.* **A47**, 431–432.
Brans, C. and Dicke, R. H.: 1961, *Phys. Rev. (Ser. 2)* **124**, 925–935.
Brecher, K.: 1977, *Phys. Rev. Lett.* **39**, 1051–1054.
Brecher, K.: 1978, *Astrophys. J.* **219**, L117–L118.
Brecher, K. and Grunsfeld, J.: 1978, *Bull. Am. Astron. Soc.* **10**, 429.
Brillouin, L.: 1971, *Relativity Reexamined*, Academic Press, New York.
Brosche, P.: 1970, *Astron. Astrophys.* **6**, 240–253.
Browne, P. F.: 1976a, *Found. Phys.* **6**, 457–471.
Browne, P. F.: 1976b, *Int. J. Theoret. Phys.* **15**, 73–79.
Bugrii, A. I. and Trushevskii, A. A.: 1977, *Astrophysics* **13**, 195–202.
Burbidge, E. M. and Burbidge, G. R.: 1954, *Phil. Mag. (Ser. 7)* **45**, 1019–1022.
Burke, K. and Kidd, W. S. F.: 1978a, *Nature* **272**, 240–241.
Burke, K. and Kidd, W. S. F.: 1978b, *Nature* **274**, 721.

Caderni, N., De Cosmo, V., Fabbri, R., Melchiorri, B., Melchiorri, F., and Natale, V.: 1977, *Phys. Rev. (Ser. 3)* **D16**, 2424–2429.
Cahill, M. and Taub, A. H.: 1971, *Comm. Math. Phys.* **21**, 1–40.
Calame, O. and Mulholland, J. D.: 1978, *Science* **199**, 977–978.
Callan, C., Coleman, S., and Jackiw, R.: 1970, *Ann. Phys. (N.Y.)* **59**, 42–73.
Caloi, V. and Firmani, C.: 1971, *Astrophys. Space Sci.* **10**, 309–327.
Campbell, J. E.: 1903, *Theory of Continuous Groups*, Oxford Un. Press, Oxford.
Canuto, V.: 1974, *Ann. Rev. Astron. Astrophys.* **12**, 167–214.
Canuto, V.: 1975, *Ann. Rev. Astron. Astrophys.* **13**, 335–380.
Canuto, V.: 1976, *Astrophys. J.* **205**, 659–673.
Canuto, V.: 1977, *Ann. Acad. Sci. (N.Y.)* **302**, 514–527.
Canuto, V.: 1978, *Monthly Notices Roy. Astron. Soc.* **184**, 721–725.
Canuto, V., Adams, P. J., Hsieh, S.-H., and Tsiang, E.: 1977a, *Phys. Rev. (Ser. 3)* **D16**, 1643–1663.
Canuto, V., Adams, P. J., Hsieh, S.-H., and Tsiang, E.: 1977b, *Nature* **265**, 763.
Canuto, V., Hsieh, S.-H., and Adams, P. J.: 1977c, *Phys. Rev. Lett.* **39**, 429–432.
Canuto, V., Datta, B., and Kalman, G.: 1978, *Astrophys. J.* **221**, 274–283.
Canuto, V. and Hsieh, S.-H.: 1977, *Astron. Astrophys.* **61**, L5–L6.
Canuto, V. and Hsieh, S.-H.: 1978a, *Astron. Astrophys.* **65**, 389–391.
Canuto, V. and Hsieh, S.-H.: 1978b, *Astrophys. J.* **224**, 302–307.
Canuto, V., Hsieh, S.-H., and Adams, P. J.: 1978, *Phys. Rev. (Ser. 3)* **D18**, 3577–3580.
Canuto, V. and Lee, J. F.: 1977, *Phys. Lett.* **B72**, 281–284.
Canuto, V. and Lodenquai, J.: 1977, *Astrophys. J.* **211**, 342–356.
Carey, S. W.: 1976, *The Expanding Earth*, Elsevier, Amsterdam.
Carmeli, M.: 1977, *Group Theory and General Relativity*, McGraw-Hill, London.
Carmeli, M., Fickler, S. I., and Witten, L. (eds.): 1970, *Relativity*, Plenum Press, London.
Carr, B. J.: 1976, *Astrophys. J.* **206**, 8–25.
Carr, B. J.: 1977a, *Monthly Notices Roy. Astron. Soc.* **181**, 293–309.
Carr, B. J.: 1977b, *Astron. Astrophys.* **56**, 377–383.
Carr, B. J.: 1978, *Comm. Astrophys.* **7**, 161–173.
Carr, B. J. and Hawking, S. W.: 1974, *Monthly Notices Roy. Astron. Soc.* **168**, 399–415.
Carr, B. J. and Rees, M.: 1977, *Astron. Astrophys.* **61**, 705–709.
Cavaliere, A., Danese, L., and de Zotti, G.: 1977a, *Astron. Astrophys.* **60**, L15–L17.
Cavaliere, A., Danese, L., and de Zotti, G.: 1977b, *Astrophys. J.* **217**, 6–15.
Cavaliere, A., Danese, L., and de Zotti, G.: 1978, *Astrophys. J.* **221**, 399–406.
Chakravarty, N., Choudhury, S. B., and Banerjee, A.: 1976, *Australian J. Phys.* **29**, 113–117.
Chandrasekhar, S.: 1931, *Monthly Notices Roy. Astron. Soc.* **91**, 456–466.
Chandrasekhar, S.: 1937, *Nature* **139**, 757–758.
Chastel, A. A.: 1976, *Astron. Astrophys.* **53**, 67–82.
Chapman, C. and Leiter, D. J.: 1976, *Am. J. Phys.* **44**, 858–862.
Chapman, S.: 1929, *Nature* **124**, 19–26.
Chapman, S.: 1948, *Monthly Notices Roy. Astron. Soc.* **108**, 236–251.
Chernyak, Y. B.: 1978, *Nature* **273**, 497–501.
Chincarini, G.: 1978a, *Nature* **272**, 515–516.
Chincarini, G.: 1978b, *Nature* **274**, 452–454.
Chincarini, G. and Rood, H. J.: 1974, *Astrophys. J.* **194**, 21–26.
Chincarini, G. and Rood, H. J.: 1975, *Nature* **257**, 294–295.
Cialdea, R.: 1972, *Cimento Nuovo (Lett., Ser. 2)* **4**, 821–825.
Clark, S. P., Turekian, K. K., and Grossman, L.: 1972, in *The Nature of the Solid Earth* (ed. E. C. Robertson), pp. 3–18, McGraw-Hill, New York.
Clemence, G. M.: 1962, Gravity Research Foundation 1st. Award Essay.

Close, F.: 1978, *Nature* **275**, 267–268.
Clube, S. V. M.: 1977, *Astrophys. Space Sci.* **50**, 425–443.
Clube, S. V. M.: 1978, *New Scientist* **80**, 284–286.
Clutton-Brock, M.: 1976, *Astrophys. Space Sci.* **41**, L9–L11.
Clutton-Brock, M.: 1977, *Astrophys. Space Sci.* **47**, 423–433.
Cohn, J.: 1975, *J. Gen. Rel. Grav.* **6**, 143–150.
Conklin, E. K.: 1969, *Nature* **222**, 971–972.
Cook, M. W.: 1975, *Australian J. Phys.* **28**, 413–422.
Corey, B. E. and Wilkinson, D. T.: 1976, *Bull. Am. Astron. Soc.* **8**, 351.
Crawford, J. F.: 1977, *Nature* **269**, 194.
Creer, K. M.: 1967, in *International Dictionary of Geophysics* (ed. S. K. Runcorn), pp. 383–389, Pergamon Press, London.
Crossely, D. J. and Stevens, R. K.: 1976, *Can. J. Earth Sci.* **13**, 1723–1724.
Cunningham, E.: 1910, *Proc. London Math. Soc.* **8**, 77–98.
Danese, L. and de Zotti, G.: 1977, *Riv. Nuovo Cimento* **7**, 277–362.
Danese, L. and de Zotti, G.: 1978, *Astron. Astrophys.* **68**, 157–164.
Datt, B.: 1938. *Z. Phys.* **108**, 314–321.
Datta, B.: 1977, Thesis, Dept. Physics, C.C.N.Y., New York.
Davies, P. C. W.: 1970, *Nature* **228**, 270–271.
Davies, P. C. W.: 1972, *Nature Phys. Sci.* **240**, 3–5.
Davies, P. C. W.: 1973, *Monthly Notices Roy. Astron. Soc.* **161**, 1–5.
Davies, P. C. W.: 1974, *The Physics of Time Asymmetry*, Surrey Univ. Press, Guildford, U.K.
Davies, P. C. W.: 1975, *J. Phys.* **A8**, 609–616.
Davies, P. C. W.: 1976a, *Monthly Notices Roy. Astron. Soc.* **177**, 179–190.
Davies, P. C. W.: 1976b, *Nature* **263**, 377–380.
Davies, P. C. W.: 1976c, *Nature* **259**, 157.
Davies, P. C. W.: 1977a, *Nature* **266**, 12–13.
Davies, P. C. W.: 1977b, *Nature* **269**, 560–561.
Davies, P. C. W.: 1977c, *Nature* **268**, 397–398.
Davies, P. C. W.: 1978a, *Nature* **273**, 268–269.
Davies, P. C. W.: 1978b, *Nature* **273**, 336–337.
Davies, P. C. W. and Taylor, J. G.: 1974, *Nature* **250**, 37–38.
Davis, M.: 1976, in *Frontiers of Astrophysics* (ed. E. H. Avrett), pp. 472–522, Harvard Univ. Press, Cambridge, Mass.
Davis, M., Groth, E. J., and Peebles, P. J. E.: 1977, *Astrophys. J.* **212**, L107–L111.
Davis, M. and Peebles, P. J. E.: 1977, *Astrophys. J. Suppl.* **34**, 425–450.
Dearnley, R.: 1966, *Phys. Chem. Earth* **7**, 1–114.
Denner, H. and Obregón, O.: 1971, *Astrophys. Space Sci.* **14**, 454–459.
Denner, H. and Obregón, O.: 1972, *Astrophys. Space Sci.* **15**, 326–333.
Delsemme, A. H. (ed.): 1977, *Comets, Asteroids, Meteorites* (I.A.U. Coll. 39), Univ. Toledo Press, Ohio.
Dermott, S. F. (ed.): 1978, *The Origin of the Solar System*, J. Wiley, New York.
Deser, S.: 1970, *Ann. Phys. (N.Y.)* **59**, 248–253.
De Vaucouleurs, G.: 1958, *Astron. J.* **63**, 253–266.
De Vaucouleurs, G.: 1961, *Astrophys. J. Suppl.* **6**, 213–234.
De Vaucouleurs, G.: 1970, *Science* **167**, 1203–1213.
De Vaucouleurs, G.: 1971, *Pub. Astron. Soc. Pac.* **83**, 113–143.
De Vaucouleurs, G.: 1972a, in *External Galaxies and Quasi-Stellar Objects* (I.A.U. Symp. 44, ed. D. S. Evans), pp. 353–366, D. Reidel, Dordrecht, Holland.
De Vaucouleurs, G.: 1972b, *Nature* **236**, 166.
De Vaucouleurs, G.: 1975a, *Astrophys. J.* **202**, 319–326.

De Vaucouleurs, G.: 1975b, *Astrophys. J.* **202**, 610—615.
De Vaucouleurs, G.: 1975c, *Astrophys. J.* **202**, 616–618.
De Vaucouleurs, G.: 1975d, in *Galaxies and the Universe* (Vol. 9, *Stars and Stellar Systems*, eds. A. Sandage, M. Sandage, and J. Kristian), pp. 557–600, Univ. Chicago Press, Chicago, Ill.
De Vaucouleurs, G.: 1976, *Astrophys. J.* **205**, 13–28.
De Vaucouleurs, G., and Corwin, H. G.: 1975, *Astrophys. J.* **202**, 327–334.
De Vaucouleurs, G., and De Vaucouleurs, A.: 1963, *Astron. J.* **68**, 96–107.
De Vaucouleurs, G., and De Vaucouleurs, A.: 1964, *Reference Catalogue of Bright Galaxies*, Texas Univ. Press, Austin, Texas.
De Vaucouleurs, G., De Vaucouleurs, A., and Corwin, H. G.: 1976, *Second Reference Catalogue of Bright Galaxies*, Texas Univ. Press, Austin, Texas.
De Vaucouleurs, G., and Peters, W. L.: 1968, *Nature* **220**, 868–874.
De Witt, B. S.: 1962, in *Gravitation: An Introduction to Current Research* (ed. L. Witten), pp. 266–381, J. Wiley, New York.
De Witt, B. and Graham, N. (eds.): 1973, *The Many Worlds Interpretation of Quantum Mechanics*, Princeton Univ. Press, Princeton, N.J.
Dicke, R. H.: 1962a, *Rev. Mod. Phys.* **34**, 110–122.
Dicke, R. H.: 1962b, *Phys. Rev. (Ser. 2)* **125**, 2163–2167.
Dicke, R. H.: 1970, *Astrophys. J.* **159**, 1–24.
Dickens, R. J., and Malin, S. R. C.: 1965, *Observatory* **85**, 260–262.
Dicus, D., Kolb, E., and Teplitz, V.: 1977, *Phys. Rev. Lett.* **39**, 168–171.
Dingle, H.: 1973, *Nature* **244**, 567–568.
Dirac, P. A. M.: 1935, *Ann. Math. (Ser. 2)* **36**, 657–669.
Dirac, P. A. M.: 1936, *Ann. Math. (Ser. 2)* **37**, 429–442.
Dirac, P. A. M.:1938, *Proc. Roy. Soc. London* **A165**, 199–208.
Dirac, P. A. M.: 1972, in *Cosmology, Fusion and Other Matters* (ed. F. Reines), pp. 56–59, A. Hilger, London.
Dirac, P. A. M.: 1973, *Proc. Roy. Soc. London* **A333**, 403–418.
Dirac, P. A. M.: 1974, in *Fundamental Theories in Physics* (eds. S. L. Mintz, L. Mittag, and S. M. Widmayer), pp. 1–18, Plenum Press, New York.
Dirac, P. A. M.: 1975, in *Theories and Experiments in High Energy Physics* (eds A. Perlmutter and S. M. Widmayer), pp. 443–455, Plenum Press, New York.
Dirac, P. A. M.: 1978, in *On the Measurement of Cosmological Variations of the Gravitational Constant* (ed. L. Halpern), pp. 3–17, Univ. Presses of Florida, Florida.
Doake, C. S. M.: 1978, *Nature* **275**, 304–305.
Dodd, R. J., Morgan, D. H., Nandy, K., Reddish, V. C., and Seddon, H.: 1975, *Monthly Notices Roy. Astron. Soc.* **171**, 329–351.
Dodd, R. J., MacGillivray, H. T., Ellis, R. S., Fong, R., and Phillipps, S.: 1976, *Monthly Notices Roy. Astron. Soc.* **176**, 33P–36P.
Domokos, G., Janson, M. M., and Kovesi-Domokos, S.: 1975, *Nature* **257**, 203–205.
Domokos, G., Janson, M. M., and Kovesi-Domokos, S.: 1976, *Nature* **259**, 157.
Doroshkevich, A. G. and Zel'dovich, Y. B.: 1975, *Astrophys. Space Sci.* **35**, 55–65.
Doroshkevich, A. S. and Shandarin, S. F.: 1976, *Monthly Notices Roy. Astron. Soc.* **175**, 15P–18P.
Drury, S. A.: 1978, *Nature* **274**, 720–721.
Dreitlein, J.: 1974, *Phys. Rev. Lett.* **33**, 1243–1244.
Dube, R. R., Wickes, W. C., and Wilkinson, D. T.: 1977, *Astrophys. J.* **215**, L51–L52.
Dyer, C. and Roeder, R. C.: 1972, *Astrophys. J.* **174**, L115–L117.
Dyer, C. and Roeder, R. C.: 1973, *Astrophys. J.* **180**, L31–L34.
Dyer, C. and Roeder, R. C.: 1974, *Astrophys. J.* **189**, 167–175.
Dyson, F. J.: 1967, *Phys. Rev. Lett.* **19**, 1291–1293.

Eddington, A.: 1936, *Relativity Theory of Protons and Electrons,* Cambridge Univ. Press, London.
Edwards, D. and Heath, D. J.: 1976, *Astrophys. Space Sci.* **41**, 183–193.
Eichler, D.: 1977, *Astrophys. J.* **218**, 579–581.
Einasto, J.: 1978, in *The Large Scale Structure of the Universe* (I.A.U. Symp. 79, eds. M. Longair and J. Einasto), pp. 51–61, D. Reidel, Dordrecht, Holland.
Einstein, A.: 1951, *The Meaning of Relativity,* Methuen, London.
Einstein, A., and Bergmann, P. G.: 1938, *Ann. Math. (Ser. 2)* **39**, 683–701.
Einstein, A., and Mayer, W.: 1931, *Sitz. Preuss. Akad. Wiss.* **25**, 541–553.
Eisenstaedt, J.: 1975a, *Phys. Rev. (Ser. 3)* **D11**, 2021–2025.
Eisenstaedt, J.: 1975b, *Phys. Rev. (Ser. 3)* **D12**, 1573–1575.
Eisenstaedt, J.: 1977, *Astrophys. J.* **211**, 16–20.
Ellis, G.: 1978a, *Comm. Astrophys.* **8**, 1–7.
Ellis, G.: 1978b, *J. Gen. Rel. Grav.* **9**, 87–94.
Ellis, G., Maartens, R., and Nel, S. D.: 1978, *Monthly Notices Roy. Astron. Soc.* **184**, 439–465.
Engstrom, H. T., and Zorn, M.: 1936, *Phys. Rev. (Ser. 2)* **49**, 701–702.
Epstein, R. I., and Petrosian, V.: 1975, *Astrophys. J.* **197**, 281–284.
Evans, A. B.: 1978, *Monthly Notices Roy. Astron. Soc.* **183**, 727–748.
Everett, H.: 1957, *Rev. Mod. Phys.* **29**, 454–462.
Fabbri, R., Melchiorri, F., and Natale, V.: 1978, *Astrophys. Space Sci.* **59**, 223–236.
Fairall, A. P.: 1978, *Monthly Notices Roy. Astron. Soc.* **183**, 59P–62P.
Fairchild, E. E.: 1977, *Astron. Astrophys.* **56**, 199–206.
Fall, S. M.: 1976, Thesis, Oxford Univ.
Fall, S. M.: 1978, *Monthly Notices Roy. Astron. Soc.* **185**, 165–177.
Fall, S. M., Geller, M. J., Jones, B. J. T., and White, S. D. M.: 1976, *Astrophys. J.* **205**, L121–L125.
Fall, S. M. and Jones, B. J. T.: 1976, *Nature* **262**, 457–460.
Fall, S. M. and Saslaw, W. C.: 1976, *Astrophys. J.* **204**, 631–641.
Fanti, C., Lari, C., and Olori, M. C.: 1978, *Astron. Astrophys.* **67**, 175–184.
Fazio, G. G., and Stecker, F. W.: 1976, *Astrophys. J.* **207**, L49–L52.
Feinberg, G.: 1969, *Science* **166**, 879–881.
Felten, J., and Morrison, P.: 1966, *Astrophys J.* **146**, 686–708.
Fennelly, A. J.: 1974, *Nature* **248**, 221–223.
Fennelly, A. J.: 1977, *Monthly Notices Roy. Astron. Soc.* **181**, 121–130.
Ferencz, C., and Tarcsai, G.: 1971, *Nature* **223**, 404–406.
Field, G. B.: 1975, in *Galaxies and the Universe* (Vol. 9, *Stars and Stellar Systems*, eds A. Sandage, M. Sandage and J. Kristian), pp. 359–407, Univ. Chicago Press, Chicago.
Field, G. B.: 1975, in *Galaxies and the Universe* (Vol. 9, *Stars and Stellar Systems*, eds. A. Sandage, M. Sandage, and J. Kristian), pp. 359–407, Univ. Chicago Press, Chicago, Ill.
Field, G. B., and Saslaw, W. C.: 1971, *Astrophys. J.* **170**, 199–206.
Finlay-Freundlich, E.: 1954a, *Phil. Mag. (Ser. 7)* **45**, 303–319.
Finlay-Freundlich, E.: 1954b, *Proc. Phys. Soc.* **A67**, 192–193.
Finzi, A.: 1962, *Phys. Rev.* **128**, 2012–2015.
Finzi, A.: 1963a, *Nuovo Cimento (Ser. 10)* **28**, 224–226.
Finzi, A.: 1963b, *Monthly Notices Roy. Astron. Soc.* **127**, 21–30.
Finzi, A.: 1968, *Icarus* **9**, 191–196.
Fischer, A. E. and Marsden, J. E.: 1973, Gravity Research Foundation 1st.·Award Essay.
Ford, L. H.: 1975, *Phys. Rev. (Ser. 3)* **D11**, 3370–3377.
Forman, M. A.: 1977, *Astrophys. J.* **212**, L1–L2.
Fox, J. G., Jacobs, K. C., Thomsen, B., and Von Hoerner, S.: 1975, *Astrophys. J.* **201**, 545–546.

Fradkin, E. S. and Tyutin, I. V.: 1974, *Riv. Nuovo Cimento* **4**, 1–78.
Franken, P. A., and Ampulski, G. W.: 1971, *Phys. Rev. Lett.* **72**, 115–117.
Freedman, D. Z., and Van Nieuwenhuizen, P.: 1978, *Sci. Am.* **238** (Feb.), 126–143.
Freeman, J. W.: 1978, in *The Origin of the Solar System* (ed. S. F. Dermott), pp. 635–640, J. Wiley, New York.
Freeman, K. C.: 1965, *Monthly Notices Roy. Astron. Soc.* **130**, 183–197.
Freund, P. G. O., Maheshwari, A., and Schonberg, E.: 1969, *Astrophys. J.* **157**, 857–867.
Freund, P. G. O.: 1974, *Ann. Phys. (N.Y.)* **84**, 440–454.
Frosch, R.: 1973, *Nuovo Cimento (Lett., Ser. 2)* **8**, 633–638.
Fry, J. N., and Peebles, P. J. E.: 1978, *Astrophys. J.* **221**, 19–33.
Gamow, G.: 1967a, *Phys. Rev. Lett.* **19**, 759–761.
Gamow, G.: 1967b, *Phys. Rev. Lett.* **19**, 913–914.
Gamow, G.: 1967c, *Proc. Nat. Acad. Sci. (U.S.A.)* **57**, 187–193.
Gamow, G.: 1967d, *Science* **158**, 766–769.
Geller, M. J., and Peebles, P. J. E.: 1972, *Astrophys. J.* **174**, 1–5.
Gerasim, A.: 1969, *Astrophys. Lett.* **4**, 51–54.
Gibbons, G.: 1976, *New Scientist* **69**, 54–56.
Gilbert, C.: 1956, *Monthly Notices Roy. Astron. Soc.* **116**, 684–690.
Gilbert, C.: 1960, *Monthly Notices Roy. Astron. Soc.* **120**, 367–386.
Gillies, G. T., Ritter, R. C., and Rood, R. T.: 1978, *Audio*, June, 76–84.
Ginzburg, V. L.: 1971, *Comments Astrophys. Space Phys.* **3**, 7–11.
Ginzburg, V. L.: 1975, *Q. J. Roy. Astron. Soc.* **16**, 265–281.
Ginzburg, V. L.: 1976, *Key Problems of Physics and Astrophysics*, M.I.R., Moscow (Engl. Transl., Imported Publications, Inc., 320 West Ohio St., Chicago, Ill.)
Ginzburg, V. L., and Frolov, V. P.: 1977, *Pis'ma v Astronomichesky Zhurnal* **2**, 474.
Ginzburg, V. L., Kirzhnits, D. A., and Lyubshin, A. A.: 1971, *Soviet Phys. JETP* **33**, 242–246.
Ginzburg, V. L., and Ozernoy, L. M.: 1977, *Astrophys. Space Sci.* **50**, 23–41.
Gitterman, M.: 1978, *Rev. Mod. Phys.* **50**, 85–106.
Gittus, J. H.: 1976, *Proc. Roy. Soc. London* **A348**, 95–99.
Gladyshev, G. P.: 1978, *Moon and Planets* **19**, 89–98.
Goded, F.: 1975a, *J. Gen. Rel. Grav.* **6**, 115–118.
Goded, F.: 1975b, *J. Gen. Rel. Grav.* **6**, 119–121.
Gold, R.: 1968, *Phys. Rev. Lett.* **20**, 219–220.
Gold, T.: 1962, *Am. J. Phys.* **30**, 403–410.
Goldman, I.: 1976, *J. Gen. Rel. Grav.* **7**, 681–685.
Goldman, I., and Rosen, N.: 1971, *J. Gen. Rel. Grav.* **2**, 367–384.
Goldman, I., and Rosen, N.: 1976, *J. Gen. Rel. Grav.* **7**, 895–901.
Goldman, I., and Rosen, N.: 1977, *Astrophys. J.* **212**, 602–604.
Goldman, I., and Rosen, N.: 1978, *Astrophys. J.* **225**, 708–711.
Goldman, S. P.: 1978, *Astrophys. J.* **226**, 1079–1086.
Goldstein, R. M.: 1969, *Science* **166**, 598–601.
Good, I. J.: 1970, *Phys. Lett.* **A33**, 383–384.
Gott, J. R.: 1977, *Ann. Rev. Astron. Astrophys.* **15**, 235–266.
Gott, J. R., Gunn, J., Schramm, D. N., and Tinsley, B.: 1974, *Astrophys. J.* **194**, 543–553.
Gott, J. R., and Rees, M.: 1975, *Astron. Astrophys.* **45**, 365–376.
Gott, J. R., and Turner, E. L.: 1976, *Astrophys. J.* **209**, 1–5.
Gott, J. R., and Turner, E. L.: 1977a, *Astrophys. J.* **213**, 309–322.
Gott, J. R., and Turner, E. L.: 1977b, *Astrophys. J.* **216**, 357–361.
Green, R. F., and Schmidt, M.: 1978, *Astrophys. J.* **220**, L1–L4.
Greenstein, G. S.: 1968, *Astrophys. Lett.* **1**, 139–141.
Gregory, S. A., and Thompson, L. A.: 1978, *Nature* **274**, 450–451.

Greisen, K.: 1966, *Phys. Rev. Lett.* **16,** 748–750.
Gribbin, J.: 1976, *Galaxy Formation – A Personal View*, Macmillan, London.
Gross, D. J., and Wess, J.: 1970, *Phys. Rev. (Ser. 3)* **D2,** 753–764.
Grossman, L.: 1972, *Geochim. Cosmochim. Acta* **36,** 597–619.
Grossman, L., and Larimer, J. W.: 1974, *Rev. Geophys. Space Phys.* **12,** 71–101.
Groth, E. J. and Peebles, P. J. E.: 1977, *Astrophys. J.* **217,** 385–405.
Groth, E. J., Peebles, P. J. E., Seldner, M., and Soneira, R. M.: 1977, *Sci. Am.* **237** (Nov), 76–98.
Gudehus, D. H.: 1978, *Nature* **275,** 514–515.
Gull, S. F., and Northover, K. J. E.: 1976, *Nature* **263,** 572–573.
Gunn, J.: 1967, *Astrophys. J.* **150,** 737–753.
Gunn, J.: 1977, *Astrophys. J.* **218,** 592–598.
Gunn, J., and Gott, J. R.: 1972, *Astrophys. J.* **176,** 1–19.
Gunn, J., Lee, B., Lerche, I., Schramm, D., and Steigman, G.: 1978, *Astrophys. J.* **223,** 1015–1031.
Gunn, J. and Tinsley, B.: 1975, *Nature* **257,** 454–457.
Gupta, S. N.: 1954, *Phys. Rev. (Ser. 2)* **96,** 1683–1685.
Gupta, S. N.: 1957, *Rev. Mod. Phys.* **29,** 334–336.
Gupta, S. N.: 1978, *Quantum Electrodynamics*, Gordon and Breach, New York.
Guthrie, B. N.: 1976, *Astrophys. Space Sci.* **43,** 425–431.
Haantjes, J.: 1940, *Proc. Kon. Akad. Wet.* **43,** 1288–1297.
Halpern, L.: 1977, *J. Gen. Rel. Grav.* **8,** 623–630.
Halpern, L. (ed.): 1978a, *On the Measurement of Cosmological Variations of the Gravitational Constant*, Univ. Presses of Florida, Florida.
Halpern, L.: 1978b, Preprint, Florida State Un., Tallahassee, Florida.
Halpern, L., and Long, C.: 1978, in *On the Measurement of Cosmological Variations of the Gravitational Constant* (ed. L. Halpern), pp. 87–89, Univ. Presses of Florida, Florida.
Hara, K.: 1974, *Publ. Astron. Soc. Japan* **26,** 299–301.
Hara, K.: 1977, *Publ. Astron. Soc. Japan* **29,** 753–763.
Harris, P.: 1969, *Can. J. Phys.* **47,** 1884–1885.
Harrison, E. R.: 1971, *Monthly Notices Roy. Astron. Soc.* **154,** 167–186.
Harrison, E. R.: 1973a, *Ann. Rev. Astron. Astrophys.* **11,** 155–186.
Harrison, E. R.: 1973b, in *Cargèse Lectures in Physics* **6** (ed. E. Schatzman), pp. 581–642, Gordon and Breach, London.
Harrison, E. R.: 1975, *Astrophys. J.* **195,** L61–L63.
Harrison, E. R.: 1977, *Am. J. Phys.* **45,** 119–124.
Hart, M. H.: 1978, *Icarus* **33,** 23–39.
Hartle, J. B. and Thorne, K. S.: 1968, *Astrophys. J.* **153,** 807–834.
Hartquist, T. W., and Cameron, A. G. W.: 1977, *Astrophys. Space Sci.* **48,** 145–158.
Harvey, A.: 1976, *J. Gen. Rel. Grav.* **7,** 891–893.
Hauser, M. G., and Peebles, P. J. E.: 1973, *Astrophys. J.* **185,** 757–785.
Hawking, S. W.: 1971, *Monthly Notices Roy. Astron. Soc.* **152,** 75–78.
Hawking, S. W.: 1974, *Nature* **248,** 30–31.
Hawking, S. W.: 1975, *Comm. Math. Phys.* **43,** 199–220.
Hawking, S. W.: 1976a, *Phys. Rev. (Ser. 3)* **D13,** 191–197.
Hawking, S. W.: 1976b, *Phys. Rev. (Ser. 3)* **D14,** 2460–2473.
Hawking, S. W.: 1977, *Sci. Am.* **236** (Jan.), 34–40.
Hawkins, G. S.: 1960, *Astron. J.* **65,** 52.
Hawkins, G. S.: 1962a, *Nuovo Cimento (Ser. 10)* **23,** 1021–1027.
Hawkins, G. S.: 1962b, *Nature* **194,** 563–564.
Hazard, C., Jauncey, D. L., Sargent, W. L. W., Baldwin, J. A., and Wampler, E. J.: 1973, *Nature* **246,** 205–208.

Heath, D. J.: 1977, *Monthly Notices Roy. Astron. Soc.* **179**, 351–358.
Heckmann, O.: 1959, *Observatory* **79**, 105–107.
Heckmann, O.: 1960, *Mitt. Hamburg Sternw.* **10**, No. 110.
Hegyi, D. and Gerber, G.: 1977, *Astrophys. J.* **218**, L7–L11.
Helfer, H. L.: 1954, *Phys. Rev. (Ser. 2)* **96**, 224–225.
Hellings, L., and Nordtvedt, K.: 1973, *Phys. Rev. (Ser. 3)* **D7**, 3593–3602.
Hellwig, H., Evenson, K. M., and Wineland, D. J.: 1978, *Phys. Today* **31** (Dec.), 23–30.
Henderson-Sellers, A., and Meadows, A. J.: 1977, *Nature* **270**, 589–591.
Henriksen, R. N. and Reinhardt, M.: 1977, *Astrophys. Space Sci.* **49**, 3–39.
Henriksen, R. N. and Wesson, P. S.: 1978a, *Astrophys. Space Sci.* **53**, 429–444.
Henriksen, R. N. and Wesson, P. S.: 1978b, *Astrophys. Space Sci.* **53**, 445–457.
Henry, P. S.: 1971, *Nature* **231**, 516–518.
Hilgenberg, O.: 1962, *Neu. Jahr. Geol. Palaeontol. (A6)* **116**, 1–56.
Hill, E. L.: 1945a, *Phys. Rev. (Ser. 2)* **67**, 358–363.
Hill, E. L.: 1945b, *Phys. Rev. (Ser. 2)* **68**, 232–233.
Hill, E. L.: 1947, *Phys. Rev. (Ser. 2)* **72**, 143–149.
Hively, R. M.: 1971, Thesis, Harvard Univ.
Hobart, R. H.: 1976, *Found. Phys.* **6**, 473–483.
Hoffmann, B.: 1947, *Phys. Rev. (Ser. 2)* **72**, 458–465.
Hoffmann, B.: 1948a, *Phys. Rev. (Ser. 2)* **73**, 30–35.
Hoffmann, B.: 1948b, *Phys. Rev. (Ser. 2)* **73**, 1042–1046.
Hoffmann, B.: 1952, *Phys. Rev. (Ser. 2)* **87**, 703–705.
Hoffmann, B.: 1953a, *Phys. Rev. (Ser. 2)* **89**, 49–52.
Hoffmann, B.: 1953b, *Phys. Rev. (Ser. 2)* **89**, 52–59.
Hofmann, W., and Lemke, D.: 1978, *Astron. Astrophys.* **68**, 389–390.
Holmberg, E. R. R.: 1956, *Monthly Notices Roy. Astron. Soc.* **116**, 691–698.
Horedt, G.: 1973, *Astrophys. J.* **183**, 383–386.
Hoyle, F.: 1949a, *Monthly Notices Roy. Astron. Soc.* **109**, 365–371.
Hoyle, F.: 1949b, *Nature* **163**, 196–198.
Hoyle, F.: 1965, *Phys. Rev. Lett.* **15**, 131–132.
Hoyle, F.: 1976, *Am. Scientist* **64** (March/April), 197–202.
Hoyle, F., and Narlikar, J. V.: 1963, *Proc. Roy. Soc. London* **A273**, 1–11.
Hoyle, F., and Narlikar, J. V.: 1968, *Nature* **219**, 340–341.
Hoyle, F., and Narlikar, J. V.: 1969, *Ann. Phys. (N.Y.)* **54**, 207–239.
Hoyle, F., and Narlikar, J. V.: 1971a, *Nature* **233**, 41–44.
Hoyle, F., and Narlikar, J. V.: 1971b, *Ann. Phys. (N.Y.)* **62**, 44–97.
Hoyle, F., and Narlikar, J. V.: 1972, *Monthly Notices Roy. Astron. Soc.* **155**, 323–335.
Hoyle, F., and Narlikar, J. V.: 1974, *Action at a Distance in Physics and Cosmology*, Freeman, San Francisco, Calif.
Hoyle, F., Wickramasinghe, N. C., and Reddish, V. C.: 1968, *Nature* **218**, 1124–1126.
Hughes, D. W.: 1978, *Nature* **273**, 489–490.
Hughes, J. L.: 1977a, in *Scientific Applications of Lunar Laser Ranging* (ed. J. D. Mulholland), pp. 289–302, D. Reidel, Dordrecht, Holland.
Hughes, J. L.: 1977b, *Astrophys. Space Sci.* **46**, L15–L18.
Hughes, V. W. and Williams, W. L.: 1969, Gravity Research Foundation 1st. Award Essay.
Hut, P., and Verhulst, F.: 1976, *Monthly Notices Roy. Astron. Soc.* **177**, 545–549.
Icke, V.: 1973, *Astron. Astrophys.* **27**, 1–21.
Infeld, L.: 1945, *Nature* **156**, 114.
Infeld, L., and Schild, A.: 1945, *Phys. Rev. (Ser. 2)* **68**, 250–272.
Infeld, L., and Schild, A.: 1946, *Phys. Rev. (Ser. 2)* **70**, 410–425.
Ingraham, R. L.: 1952, *Proc. Nat. Acad. Sci. (U.S.A.)* **38**, 921–924.

Isaak, G. R.: 1969, *Nature* **223**, 161.
Isham, C. J., Penrose, R., and Sciama, D. W. (eds.): 1975, *Quantum Gravity: An Oxford Symposium*, Oxford Univ. Press, Oxford.
Isham, C. J., Salam, A., and Strathdee, J.: 1971, *Phys. Rev. (Ser. 3)* **D3**, 867–873.
Jaakkola, T.: 1971, *Nature* **234**, 534–535.
Jaakkola, T., Teerikorpi, P., and Donner, K. J.: 1975a, *Astron. Astrophys.* **40**, 257–266.
Jaakkola, T., Karoji, H., Moles, M., and Vigier, J.-P.: 1975b, *Nature* **256**, 24–25.
Jaakkola, T., Karoji, H., Le Denmat, G., Moles, M., Nottale, L., Vigier, J.-P., and Pecker, J.-C.: 1976, *Monthly Notices Roy. Astron. Soc.* **177**, 191–213.
Jaakkola, T., and Moles, M.: 1976, *Astron. Astrophys.* **53**, 389–396.
Jaakkola, T., Moles, M., and Vigier, J.-P.: 1978, *Astrophys. Space Sci.* **58**, 99–112.
Jacobs, J. A.: 1970, *Phys. Earth Planet Interiors* **2**, 303–310.
Jaki, S. L.: 1977, in *Cosmology, History and Theology* (eds. W. Yourgrau and A. D. Breck), pp. 233–251, Plenum Press, New York.
Jánossy, L.: 1971, *Theory of Relativity Based on Physical Reality*, Akademiai Kiadó, Budapest.
Jastrow, R., and Cameron, A. G. W. (eds.): 1963, *Origin of the Solar System*, Academic Press, New York.
Jauncey, D. L. (ed.): 1977, *Radio Astronomy and Cosmology* (I.A.U. Symp. 74), D. Reidel, Dordrecht, Holland.
Jeffreys, H.: 1970, *The Earth* (5th ed.), Cambridge Univ. Press, London.
Jones, B. J. T.: 1973, *Astrophys. J.* **181**, 269–294.
Jones, B. J. T.: 1976, *Rev. Mod. Phys.* **48**, 107–149.
Jones, B. J. T.: 1977, *Monthly Notices Roy. Astron. Soc.* **180**, 151–162.
Jordan, P., 1949, *Nature* **164**, 637–640.
Jordan, P.: 1962, *Rev. Mod. Phys.* **34**, 596–600.
Just, K.: 1959, *Astrophys. J.* **129**, 268–270.
Kahn, P. and Pompea, S.: 1978, *Nature* **275**, 606–611.
Kalitzin, N. St.: 1967, *Liège Symposia* **15**, 81–123, Cointe Sclessin, Belgium.
Kaluza, T.: 1921, *Sitz. Preuss. Akad. Wiss.* **54**, 966–975.
Kanasewich, E. R.: 1976, *Can. J. Earth Sci.* **13**, 331–339.
Kantor, W., and Szekeres, G.: 1956, *Phys. Rev. (Ser. 2)* **104**, 831–834.
Kantowski, R.: 1969, *Astrophys. J.* **155**, 1023–1027.
Kapp, R. O.: 1960, *Towards a Unified Cosmology*, Hutchinson, London.
Karachentsev, I. D.: 1978, in *The Large Scale Structure of the Universe* (I.A.U. Symp. 79, eds. M. Longair and J. Einasto), pp. 11–20, D. Reidel, Dordrecht, Holland.
Karlsson, K.: 1971, *Astron. Astrophys.* **13**, 333–335.
Karlsson, K.: 1977, *Astron. Astrophys.* **58**, 237–240.
Karoji, H.: 1978, *Astron. Astrophys.* **69**, 375–380.
Kartaschoff, P.: 1978, *Frequency and Time*, Academic Press, London.
Katz, J.: 1974, *Int. J. Theoret. Phys.* **10**, 165–173.
Kaula, W. M.: 1975, *Icarus* **26**, 1–15.
Kazanas, D., Schramm, D. N., and Hainebach, K.: 1978, *Nature* **274**, 672–673.
Kellogg, E. M.: 1977, *Astrophys. J.* **218**, 582–591.
Kermack, W. O., and McCrea, W. H.: 1933, *Monthly Notices Roy. Astron. Soc.* **93**, 519–529.
Kharbediya, L. I.: 1976, *Soviet Astron.* **20**, 647–650.
Kharbediya, L. I.: 1977, *Soviet Astron.* **21**, 672–674.
Kiang, T.: 1967, *Monthly Notices Roy. Astron. Soc.* **135**, 1–22.
Kilmister, C. W.: 1966, *Sir Arthur Eddington*, pp. 269–271, Pergamon Press, Oxford.
King, J. G.: 1960, *Phys. Rev. Lett.* **5**, 562–565.
King, L. A.: 1970, *New Scientist* **45**, 127.

King, R. W., Counselman, C. C., and Shapiro, I.: 1978, *J. Geophys. Res.* **83**, 3377–3381.
Kirzhnits, D. A.: 1972, *Soviet Phys. JETP, Lett.* **15**, 529–531.
Kirzhnits, D. A., and Linde, A. D.: 1972, *Phys. Lett.* **B42**, 471–474.
Kirzhnits, D. A., and Linde, A. D.: 1975, *Soviet Phys. JETP* **40**, 628–634.
Kirzhnits, D. A., and Linde, A. D.: 1976, *Ann. Phys. (N.Y.)* **101**, 195–238.
Klein, O.: 1926, *Z. Phys.* **37**, 895–906.
Klein, O.: 1927, *Z. Phys.* **46**, 188–208.
Knauth, L. P., and Epstein, S.: 1976, *Geochim. Cosmochim. Acta* **40**, 1095–1108.
Köhler, E.: 1978, *Astron. Astrophys.* **70**, 163–164.
Krat, V. A., and Gerlovin, L.: 1974, *Astrophys. Space Sci.* **26**, 521–522.
Krat, V. A., and Gerlovin, L.: 1975a, *Astrophys. Space Sci.* **33**, L5–L8.
Krat, V. A., and Gerlovin, L.: 1975b, *Astrophys. Space Sci.* **34**, L11–L12.
Kuchař, K.: 1973, in *Relativity, Astrophysics and Cosmology* (Proc. Banff Summer School), pp. 238–283, D. Reidel, Dordrecht, Holland.
Kuhi, L. V., Pecker, J.-C., and Vigier, J.-P.: 1974, *Astron. Astrophys.* **32**, 111–114.
Kundt, W.: 1976, *Nature* **259**, 30–31.
Kursunoglu, B.: 1952, *Phys. Rev. (Ser. 2)* **88**, 1369–1379.
Kursunoglu, B.: 1957, *Rev. Mod. Phys.* **29**, 412–416.
Kursunoglu, B.: 1960, *Nuovo Cimento (Ser. 10)* **15**, 729–756.
Kursunoglu, B.: 1974a, in *Fundamental Theories in Physics* (eds. S. L. Mintz, L. Mittag, and S. M. Widmayer), pp. 19–20, 21–120, Plenum Press, New York.
Kursunoglu, B.: 1974b, *Phys. Rev. (Ser. 3)* **D9**, 2723–2745.
Lake, G., and Partridge, R. B.: 1977, *Nature* **270**, 502.
Larson, R.: 1976, in *Galaxies* (6th. Adv. Course of the Swiss Soc. of Astronomy and Astrophysics, Saas-Fee, eds. L. Martinet and M. Mayor), pp. 67–154, Geneva Obs., Switzerland.
Layzer, D.: 1954, *Astron. J.* **59**, 268–270.
Layzer, D.: 1968a, *Astrophys. Lett.* **1**, 99–102.
Layzer, D.: 1968b, Gravity Research Foundation 1st. Award Essay.
Layzer, D.: 1971, in *Astrophysics and General Relativity*, Vol. 2 (Brandeis Un. Summer Institute in Theoretical Physics, eds. M. Chrétien, S. Deser, and J. Goldstein), pp. 151–233, Gordon and Breach, New York.
Layzer, D.: 1975, in *Galaxies and the Universe* (Vol. 9, *Stars and Stellar Systems*, eds. A. Sandage, M. Sandage, and J. Kristian), pp. 665–723, Univ. Chicago Press, Chicago.
Layzer, D., and Hively, R.: 1973, *Astrophys. J.* **179**, 361–369.
Leader, E., and Williams, P. G.: 1975, *Nature* **257**, 93–99.
Le Denmat, G., Moles, M., Vigier, J.-P., and Nieto, J.-L.: 1975, *Nature* **257**, 773–774.
Lee, B., and Weinberg, S.: 1977, *Phys. Rev. Lett.* **39**, 165–168.
Lee, D. L., Caves, C. M., Ni, W.-T., and Will, C. M.: 1976, *Astrophys. J.* **206**, 555–558.
Lee, D. L., Lightman, A. P., and Ni, W.-T.: 1974, *Phys. Rev. (Ser. 3)* **D10**, 1685–1700.
Lee, D. T.: 1974, *Phys. Rev. (Ser. 3)* **D10**, 2374–2383.
Leibowitz, E., and Rosen, N.: 1973, *J. Gen. Rel. Grav.* **4**, 449–474.
Lemaître, G.: 1961, *Astron. J.* **66**, 603–606.
Lense, J., and Thirring, H.: 1918, *Phys. Zeits.* **19**, 156–163.
Lenz, F.: 1951, *Phys. Rev. (Ser. 2)* **82**, 554.
Lerche, I.: 1978a, *Astrophys. Space Sci.* **53**, 21–37.
Lerche, I.: 1978b, *Astrophys. Space Sci.* **53**, 39–54.
Lessner, G.: 1974, *Astrophys. Space Sci.* **30**, L5–L7.
Levy, E. H.: 1978, *Nature* **276**, 481.
Levy, G. S., Sato, T., Seidel, B. L., Stelzreid, C. T., Ohlson, J. E., and Rusch, W. V. T.: 1969, *Science* **166**, 596–598.

Lewis, B. M.: 1969, Thesis, Australian National Univ.
Lewis, B. M.: 1971, *Nature Phys. Sci.* **230**, 13–15.
Lewis, B. M.: 1975, *Observatory* **95**, 168–171.
Lewis, B. M.: 1976, *Nature* **261**, 302–304.
Lewis, B. M.: 1977, *Proc. Astron. Soc. Aust.* **3**, 140–142.
Lewis, J. S.: 1972a, *Earth Planet. Sci. Lett.* **15**, 266–290.
Lewis, J. S.: 1972b, *Icarus* **16**, 241–252.
Lewis, J. S.: 1973, *Ann. Rev. Phys. Chem.* **24**, 339–351.
Liang, E. P.: 1976, *Astrophys. J.* **204**, 235–250.
Liang, E. P.: 1977, *Monthly Notices Roy. Astron. Soc.* **180**, 117–123.
Lichnerowicz, A.: 1955, *Théories Relativistes de la Gravitation et de l'Électromagnétisme*, Masson et Cie, Paris.
Lie, S.: 1893, *Theorie der Transformationsgruppen*, Teubner, Leipzig.
Lillie, C. F.: 1972, *Scientific Results from the OAO-2*, p. 583.
Lin, D. N., Carr, B. J., and Fall, S. M.: 1976, *Monthly Notices Roy. Astron. Soc.* **177**, 51–64.
Linde, A. D.: 1974, *Soviet Phys. JETP, Lett.* **19**, 183–184.
Long, D. R.: 1976, *Nature* **260**, 417–418.
Longair, M.: 1966, *Monthly Notices Roy. Astron. Soc.* **133**, 421–436.
Longair, M. (ed.): 1974, *Confrontation of Cosmological Theories with Observational Data* (I.A.U. Symp. 63), D. Reidel, Dordrecht, Holland.
Longair, M.: 1976, *Q. J. Roy. Astron. Soc.* **17**, 422–447.
Longair, M.: 1978, in *Observational Cosmology* (8th. Adv. Course of the Swiss Soc. of Astronomy and Astrophysics, Saas-Fee, eds. A. Maeder, L. Martinet, and G. Tammann), pp. 125–257, Geneva Obs., Switzerland.
Longair, M., and Einasto, J. (eds.): 1978, *The Large Scale Structure of the Universe* (I.A.U. Symp. 79), D. Reidel, Dordrecht, Holland.
Lubkin, G. B.: 1971, *Physics Today* **24** (Aug.), 17–19.
Lynden-Bell, D., and Pineault, S.: 1978a, *Monthly Notices Roy. Astron. Soc.* **185**, 679–694.
Lynden-Bell, D., and Pineault, S.: 1978b, *Monthly Notices Roy. Astron. Soc.* **185**, 695–712.
Lyttleton, R. A.: 1963a, *Proc. Roy. Soc. London* **A275**, 1–22.
Lyttleton, R. A.: 1963b, Tech. Rep. 35/522, Jet. Prop. Lab., Pasadena, Calif.
Lyttleton, R. A.: 1965a, *Proc. Roy. Soc. London* **A287**, 471–493.
Lyttleton, R. A.: 1965b, *Monthly Notices Roy. Astron. Soc.* **129**, 21–39.
Lyttleton, R. A.: 1965c, *Monthly Notices Roy. Astron. Soc.* **130**, 95–96.
Lyttleton, R. A.: 1969, *Astrophys. Space Sci.* **5**, 18–35.
Lyttleton, R. A.: 1970, *Adv. Astron. Astrophys.* **7**, 83–145.
Lyttleton, R. A.: 1973, *Moon* **7**, 422–439.
Lyttleton, R. A.: 1976, *Moon* **16**, 41–58.
Lyttleton, R. A.: 1977, *Astrophys. Space Sci.* **49**, L1–L6.
Lyttleton, R. A.: 1978a, *Moon and Planets* **19**, 101–104.
Lyttleton, R. A.: 1978b, *Moon and Planets* **19**, 425–442.
Lyttleton, R. A., and Bondi, H.: 1959, *Proc. Roy. Soc. London* **A252**, 313–333.
Lyttleton, R. A., and Fitch, J. P.: 1977, *Monthly Notices Roy. Astron. Soc.* **180**, 471–477.
Lyttleton, R. A., and Fitch, J. P.: 1978a, *Astrophys. J.* **221**, 412–413.
Lyttleton, R. A., and Fitch, J. P.: 1978b, *Moon and Planets* **18**, 233–240.
Machalski, J., Zieba, S., and Maslowski, J.: 1974, *Astron. Astrophys.* **33**, 357–361.
Maeder, A.: 1974, *Astron. Astrophys.* **32**, 177–190.
Maeder, A.: 1975, *Astron. Astrophys.* **43**, 61–69.
Maeder, A.: 1976, *Astron. Astrophys.* **47**, 389–400.
Maeder, A.: 1977a, *Astron. Astrophys.* **56**, 359–367.
Maeder, A.: 1977b, *Astron. Astrophys.* **57**, 125–133.

Maeder, A.: 1978a, *Astron. Astrophys.* **65**, 337–343.
Maeder, A.: 1978b, *Astron. Astrophys.* **67**, 81–86.
Magnenat, P., Martinet, L., and Maeder, A.: 1978, *Astron. Astrophys.* **67**, 51–53.
Maiti, S. R.: 1978, *Monthly Notices Roy. Astron. Soc.* **185**, 293–295.
Malin, S.: 1974, *Phys. Rev. (Ser. 3)* **D9**, 3228–3234.
Mansfield, V. N.: 1976, *Astrophys. J.* **210**, L137–L138.
Mansinha, L., Smylie, D. E., and Beck, A. E. (eds.): 1970, *Earthquake Displacement Fields and the Rotation of the Earth*, D. Reidel, Dordrecht, Holland.
Marchant, A., and Mansfield, V. N.: 1977, *Nature* **270**, 699–700.
Markov, M. A.: 1967, *Soviet Phys. JETP* **24**, 584–592.
Markowitz, W.: 1968, *Science* **162**, 1387–1388.
Maslowski, J., Machalski, J., and Zieba, S.: 1973, *Astron. Astrophys.* **28**, 289–294.
Masson, C. R.: 1978, *Monthly Notices Roy. Astron. Soc.* **185**, 9P–13P.
Matilla, K.: 1976, *Astron. Astrophys.* **47**, 77–95.
Matsuda, T.: 1972, *Prog. Theor. Phys.* **48**, 341–343.
Matzner, R., Ryan, M., and Toton, E.: 1973, *Nuovo Cimento (Ser. 11)* **B14**, 161–172.
Maucherat-Joubert, M., Cruvellier, P., and Deharveng, J. M.: 1978, *Astron. Astrophys.* **70**, 467–472.
MacCallum, M. A. H.: 1973, in *Cargèse Lectures in Physics* **6** (ed. E. Schatzman), pp. 61–174, Gordon and Breach, London.
MacCallum, M.: 1977, *Nature* **269**, 201–202.
McClelland, J., and Silk, J.: 1977a, *Astrophys. J.* **216**, 665–681.
McClelland, J., and Silk, J.: 1977b, *Astrophys. J.* **217**, 331–352.
McClelland, J., and Silk, J.: 1978, *Astrophys. J. Suppl.* **36**, 389–404.
McCrea, W. H.: 1950, *Endeavor* **9**, 3–10.
McCrea, W. H.: 1951, *Proc. Roy. Soc. London* **A206**, 562–575.
McCrea, W. H.: 1954, *Phil. Mag. (Ser. 7)* **45**, 1010–1018.
McCrea, W. H.: 1973, *Nature* **244**, 537.
McCrea, W. H.: 1978, *Observatory* **98**, 52–54.
McElhinny, M. W., Taylor, S. R., and Stephenson, D. J.: 1978, *Nature* **271**, 316–321.
McGruder, C. H.: 1978, *Nature* **272**, 806–807.
McVittie, G. C.: 1933, *Monthly Notices Roy. Astron. Soc.* **93**, 325–339.
McVittie, G. C.: 1940, *Observatory* **63**, 273–281.
McVittie, G. C.: 1941, *Observatory* **64**, 23–25.
McVittie, G. C.: 1942, *Proc. Roy. Soc. Edinburgh* **A61**, 210–222.
McVittie, G. C.: 1945, *Proc. Roy. Soc. Edinburgh* **A62**, 147–155.
McVittie, G. C.: 1965, *General Relativity and Cosmology* (2nd. ed.), pp. 33, 77, Chapman and Hall, London.
McVittie, G. C.: 1967, *Ann. Inst. H. Poincaré* **6**, 1–15.
McVittie, G. C.: 1978, *Monthly Notices Roy. Astron. Soc.* **183**, 749–764.
McVittie, G. C., and Stabell, R.: 1967, *Ann. Inst. H. Poincaré* **7**, 103–114.
McVittie, G. C., and Stabell, R.: 1968, *Ann. Inst. H. Poincaré* **9**, 371–393.
Melvin, M. A.: 1955, *Phys. Rev. (Ser. 2)* **98**, 884–887.
Merat, P., Pecker, J.-C., and Vigier, J.-P.: 1974, *Astron. Astrophys.* **30**, 167–174.
Mészáros, P.: 1974, *Astron. Astrophys.* **37**, 225–228.
Mészáros, P.: 1975, *Astron. Astrophys.* **38**, 5–13.
Miles, M. K., and Gildersleeves, P. B.: 1978, *Nature* **271**, 735–736.
Miller, R. H.: 1978, *Astrophys. J.* **224**, 32–38.
Milne, E. A.: 1933a, *Z. Astrophys.* **6**, 1–95.
Milne, E. A.: 1933b, *Z. Astrophys.* **7**, 180–187.
Milne, E. A.: 1935, *Relativity, Gravitation and World Structure*, Oxford Univ. Press, London.

Milne, E. A.: 1940, *Astrophys. J.* **91,** 129–158.
Milne, E. A.: 1941, *Observatory* **64,** 11–16.
Milne, E. A.: 1948, *Kinematic Relativity*, Oxford Univ. Press, London.
Milne, E. A., and Whitrow, G. J.: 1938, *Z. Astrophys.* **15,** 263–298.
Misner, C. W.: 1968, *Astrophys. J.* **151,** 431–457.
Moles, M., and Nottale, L.: 1978, *Astron. Astrophys.* **70,** 13–17.
Monnet, G., and Deharveng, J. M.: 1977, *Astron. Astrophys.* **58,** L1–L3.
Morganstern, R. E.: 1971a, *Phys. Rev. (Ser. 3)* **D4,** 278–282.
Morganstern, R. E.: 1971b, *Phys. Rev. (Ser. 3)* **D4,** 282–286.
Morganstern, R. E.: 1971c, *Nature* **232,** 109–110.
Morganstern, R. E.: 1971d, *Phys. Rev. (Ser. 3)* 2946–2950.
Morganstern, R. E.: 1972, *Nature Phys. Sci.* **237,** 70–71.
Morganstern, R. E.: 1973, *Phys. Rev. (Ser. 3)* **D7,** 1570–1579.
Morss, D. A., and Kuhn, W. R.: 1978, *Icarus* **33,** 40–49.
Motz, L.: 1953, *Phys. Rev. (Ser. 2)* **89,** 60–66.
Motz, L.: 1960, *Phys. Rev. (Ser. 2)* **119,** 1102–1105.
Motz, L.: 1961, *Nature* **189,** 994–995.
Motz, L.: 1962, *Nuovo Cimento (Ser. 10)* **26,** 672–697.
Motz, L.: 1970, *Nuovo Cimento (Ser. 10)* **A65,** 326–332.
Motz, L.: 1972, *Nuovo Cimento (Ser. 11)* **B12,** 239–255.
Mulholland, J. D. (ed.): 1977, *Scientific Applications of Lunar Laser Ranging*, D. Reidel, Dordrecht, Holland.
Muller, P. M.: 1975, Thesis, Newcastle Univ., U.K.
Muller, P. M.: 1978, in *On the Measurement of Cosmological Variations of the Gravitational Constant* (ed. L. Halpern), pp. 91–116, Univ. Presses of Florida, Florida.
Muller, R. A.: 1978, *Sci. Am.* **238** (May), 64–74.
Nagata, T.: 1970, *Phys. Earth Planet. Interiors* **2,** 311–317.
Nariai, H., and Ueno, Y.: 1960a, *Prog. Theor. Phys.* **24,** 593–613.
Nariai, H., and Ueno, Y.: 1960b, *Prog. Theor. Phys.* **24,** 1149–1165.
Nelson, A. H.: 1972, *Monthly Notices Roy. Astron. Soc.* **158,** 159–175.
Ness, N. F.: 1978, *Moon and Planets* **18,** 427–439.
Netzer, H., Yahil, A., and Yaniv, R.: 1978, *Monthly Notices Roy. Astron. Soc.* **184,** 21P–25P.
Newman, M. J., and Rood, R. T.: 1978, *Science* **198,** 1035–1037.
Newman, R.: 1977, Report at 8th. International Conference on General Relativity and Gravitation, 7–12 Aug., Waterloo Univ., Ontario, Canada.
Newman, W. I.: 1977, *Astrophys. Space Sci.* **47,** 99–108.
Newton, R. R.: 1972, *Astrophys. Space Sci.* **16,** 179–200.
Newton, R. R.: 1977, *The Crime of Claudius Ptolemy*, Johns Hopkins Univ. Press, Balt.
Ni, W.-T.: 1972, *Astrophys. J.* **176,** 769–796.
Ni, W.-T.: 1973, *Phys. Rev. (Ser. 3)* **D7,** 2880–2883.
Nickerson, J. C.: 1975, *Int. J. Theor. Phys.* **14,** 367–372.
Nicoll, J. F., and Segal, I.: 1975, *Proc. Nat. Acad. Sci. (U.S.A.)* **72,** 4691–4695.
Nicoll, J. F., and Segal, I.: 1978, *Ann. Phys. (N.Y.)* **113,** 1–28.
Nieto, J.-L.: 1977, *Nature* **270,** 411–412.
Nieto, J.-L.: 1978, *Astron. Astrophys.* **70,** 219–226.
Noerdlinger, P. D.: 1968, *Phys. Rev. (Ser. 2)* **170,** 1175.
Noerdlinger, P. D.: 1970, *Astrophys. J.* **159,** L179–L183.
Noerdlinger, P. D.: 1977, *Astrophys. J.* **218,** 317–322.
Noerdlinger, P. D.: 1978, *Astrophys. J.* **220,** 373–375.
Nordtvedt, K.: 1968, *Phys. Rev. (Ser. 2)* **169,** 1017–1023.
Nordtvedt, K.: 1969, *Phys. Rev. (Ser. 2)* **180,** 1293–1298.

Nordtvedt, K.: 1970a, *Int. J. Theor. Phys.* **3**, 133–139.
Nordtvedt, K.: 1970b, *Astrophys. J.* **161**, 1059–1067.
Nordtvedt, K.: 1971, *Phys. Rev. (Ser. 3)* **D3**, 1683–1689.
Nordtvedt, K.: 1977, in *Proc. 1st. Marcel Grossmann Meeting on General Relativity* (ed. R. Ruffini), pp. 539–544, North-Holland, Amsterdam.
Nordtvedt, K., and Will, C. M.: 1972, *Astrophys. J.* **177**, 775–792.
Norman, C. A., and Silk, J.: 1978, *Astrophys. J.* **224**, 293–301.
Nottale, L., and Moles, M.: 1978, *Astron. Astrophys.* **66**, 355–358.
Novello, M., and Rebouças, M. J.: 1978, *Astrophys. J.* **225**, 719–724.
Obregón, O., and Chauvet, P.: 1978, *Astrophys. Space Sci.* **56**, 335–339.
Occhionero, F., and Vagnetti, F.: 1975, *Astron. Astrophys.* **44**, 329–333.
Olson, D. W.: 1978, *Astrophys. J.* **219**, 777–780.
Olson, D. W., and Silk, J.: 1978, *Astrophys. J.* **226**, 50–54.
Omote, M.: 1971, *Nuovo Cimento (Lett., Ser. 2)* **2**, 58–60.
Omote, M.: 1974, *Nuovo Cimento (Lett., Ser. 2)* **10**, 33–37.
Osmer, P. S., and Smith, M. G.: 1977, *Astrophys. J.* **217**, L73–L76.
Oster, L., and Philip, K. W.: 1961, *Nature* **189**, 43.
Ostriker, J., and Peebles, P. J. E.: 1973, *Astrophys. J.* **186**, 407–480.
Ostriker, J., Peebles, P. J. E., and Yahil, A.: 1974, *Astrophys. J.* **193**, L1–L4.
Ostriker, J., and Thuan, T. X.: 1975, *Astrophys. J.* **202**, 353–365.
Ovenden, M. W., and Byl, J.: 1976, *Astrophys. J.* **206**, 57–65.
Oversby, V. M., and Ringwood, A. E.: 1971, *Nature* **234**, 463–465.
Owen, H. G.: 1976, *Phil. Tran. Roy. Soc. London* **A281**, 223–291.
Owen, P. R.: 1977, *Astrophys. Space Sci.* **49**, L7–L9.
Ozernoi, L. M.: 1972, *Soviet Astron.* **15**, 923–933.
Pachner, J.: 1965, *Monthly Notices Roy. Astron. Soc.* **131**, 173–176.
Page, L.: 1936a, *Phys. Rev. (Ser. 2)* **49**, 254–268.
Page, L.: 1936b, *Phys. Rev. (Ser. 2)* **49**, 946.
Page, L., and Adams, N. I.: 1936a, *Phys. Rev. (Ser. 2)* **49**, 466–469.
Page, L., and Adams, N. I.: 1936b, *Phys. Rev. (Ser. 2)* **49**, 703.
Pagel, B. E. J.: 1977, *Monthly Notices Roy. Astron. Soc.* **179**, 81P–85P.
Parker, L.: 1976, *Nature* **261**, 20–23.
Parkin, D. W.: 1978, *Nature* **276**, 323–324.
Pavšić, M.: 1975, *Int. J. Theor. Phys.* **14**, 299–311.
Partridge, R. B.: 1973, *Nature* **244**, 263–265.
Partridge, R. B., and Peebles, P. J. E.: 1967a, *Astrophys. J.* **147**, 868–886.
Partridge, R. B., and Peebles, P. J. E.: 1967b, *Astrophys. J.* **148**, 377–397.
Partridge, R. B., and Wilkinson, D.: 1967, *Phys. Rev. Lett.* **18**, 557–559.
Pauli, W.: 1933, *Ann. der Phys. (Ser. 5)* **18**, 305–336.
Pecker, J.-C., Roberts, A. P., and Vigier, J.-P.: 1972, *Nature* **237**, 227–229.
Pecker, J.-C., Tait, W., and Vigier, J.-P.: 1973, *Nature* **241**, 338–340.
Peebles, P. J. E.: 1967, *Astrophys. J.* **147**, 859–863.
Peebles, P. J. E.: 1970, *Astron. J.* **75**, 13–20.
Peebles, P. J. E.: 1971a, *Physical Cosmology*, Princeton Univ. Press, Princeton, N.J.
Peebles, P. J. E.: 1971b, *Comm. Astrophys. Space Phys.* **3**, 20–26.
Peebles, P. J. E.: 1973, *Astrophys. J.* **185**, 413–440.
Peebles, P. J. E.: 1974a, *Astrophys. J. Suppl.* **28**, 37–50.
Peebles, P. J. E.: 1974b, *Astrophys. J.* **189**, L51–L53.
Peebles, P. J. E.: 1974c, *Astron. Astrophys.* **32**, 197–202.
Peebles, P. J. E.: 1974d, *Astron. Astrophys.* **32**, 391–397.
Peebles, P. J. E.: 1974e, *Astrophys. Space Sci.* **31**, 403–410.

Peebles, P. J. E.: 1975, *Astrophys. J.* **196**, 647–652.
Peebles, P. J. E.: 1978a, *Comm. Astrophys.* **7**, 197–205.
Peebles, P. J. E.: 1978b, in *The Large Scale Structure of the Universe* (I.A.U. Symp. 79, eds. M. Longair and J. Einasto), pp. 217–226, D. Reidel, Dordrecht, Holland.
Peebles, P. J. E.: 1978c, *Astron. Astrophys.* **68**, 345–352.
Peebles, P. J. E., and Dicke, R. H.: 1962a, *J. Geophys. Res.* **67**, 4063–4070.
Peebles, P. J. E., and Dicke, R. H.: 1962b, *Phys. Rev. (Ser. 2)* **128**, 2006–2011.
Peebles, P. J. E., and Groth, E. J.: 1975, *Astrophys. J.* **196**, 1–11.
Peebles, P. J. E., and Hauser, M. G.: 1974, *Astrophys. J. Suppl.* **28**, 19–36.
Peebles, P. J. E., and Yu, J. T.: 1970, *Astrophys. J.* **162**, 815–836.
Pegg, D. T.: 1971, *Monthly Notices Roy. Astron. Soc.* **154**, 321–327.
Pegg, D. T.: 1973, *Nature Phys. Sci.* **246**, 40–41.
Pegg, D. T.: 1977, *Nature* **267**, 408–409.
Perl, M. L.: 1978, *Nature* **275**, 273–278.
Perrenod, S. C.: 1978, *Astrophys. J.* **224**, 285–292.
Petrosian, V., and Saltpeter, E. E.: 1968, *Astrophys. J.* **151**, 411–429.
Phillipps, S., Fong, R., Ellis, R. S., Fall, S. M. and MacGillivray, H. T.: 1978, *Monthly Notices Roy. Astron. Soc.* **182**, 673–685.
Piddington, J. H.: 1974, *Astrophys. Space Sci.* **31**, 225–240.
Pietenpol, J. L., Incoul, R., and Speiser, D.: 1974, *Phys. Rev. Lett.* **33**, 387–388.
Plagemann, S. H.: 1973, *Monthly Notices Roy. Astron. Soc.* **164**, 303–319.
Pollaine, S.: 1978, *Nature* **271**, 426–427.
Ponnamperuma, C. (ed.): 1976, *Chemical Evolution of the Giant Planets*, Academic Press, New York.
Prakash, A.: 1978, *Bull. Am. Astron. Soc.* **10**, 429.
Press, W. H., and Schechter, P.: 1974, *Astrophys. J.* **187**, 425–438.
Prokhovnik, S. J.: 1964, *Proc. Camb. Phil. Soc.* **60**, 265–271.
Prokhovnik, S. J.: 1967, *The Logic of Special Relativity*, Cambridge Univ. Press, London.
Prokhovnik, S. J.: 1968, *Int. J. Theor. Phys.* **1**, 101–105.
Provhovnik, S. J.: 1970a, *Nature* **225**, 359–361.
Prokhovnik, S. J.: 1970b, *Proc. Camb. Phil. Soc.* **67**, 391–395.
Rainey, G. W.: 1977, Thesis, California Univ.
Ramsey, W. H.: 1949, *Monthly Notices Roy. Astron. Soc., Geophys. Suppl.* **5**, 409–426.
Ramsey, W. H.: 1950, *Monthly Notices Roy. Astron. Soc., Geophys. Suppl.* **6**, 42–49.
Rastall, P.: 1968, *Can. J. Phys.* **46**, 2155–2179.
Rastall, P.: 1975, *Am. J. Phys.* **43**, 591–595.
Rastall, P.: 1976, *Can. J. Phys.* **54**, 66–75.
Rastall, P.: 1977a, *Can. J. Phys.* **55**, 38–42.
Rastall, P.: 1977b, *Can. J. Phys.* **55**, 1342–1348.
Rastall, P.: 1977c, *Astrophys. J.* **213**, 234–238.
Rastall, P.: 1978, *Astrophys. J.* **220**, 745–748.
Reasenberg, R. D., and Shapiro, I.: 1976, in *Atomic Masses and Fundamental Constants* (eds. J. H. Sanders and A. H. Wapstra), pp. 643–649, Plenum Press, New York.
Reasenberg, R. D., and Shapiro, I.: 1978 in *On the Measurement of Cosmological Variations of the Gravitational Constant* (ed. L. Halpern), pp. 71–86, Univ. Presses of Florida, Florida.
Rees, M.: 1971, in *Nuclei of Galaxies* (Vatican Study Week, ed. D. J. K. O'Connell), pp. 633–651, North-Holland Pub. Co. Amsterdam.
Rees, M.: 1972a, *Phys. Rev. Lett.* **28**, 1669–1671.
Rees, M.: 1972b, in *External Galaxies and Quasi-Stellar Objects* (I.A.U. Symp. 44, ed. D. S. Evans), pp. 407–436, D. Reidel, Dordrecht, Holland.
Rees, M.: 1977a, *Q. J. Roy. Astron. Soc.* **18**, 429–442.

Rees, M.: 1977b, in *The Evolution of Galaxies and Stellar Populations* (eds. B. Tinsley and R. Larson), pp. 339–360, Yale Univ. Obs. New Haven, Conn.
Rees, M.: 1978a, *Nature* **275**, 35–37.
Rees, M.: 1978b, *Physica Scripta* **17**, 371–376.
Rees, M.: 1978c, in *Observational Cosmology* (8th. Adv. Course of the Swiss Soc. of Astronomy and Astrophysics, Saas-Fee, eds. A. Maeder, L. Martinet and G. Tammann), pp. 259–321, Geneva Obs., Switzerland.
Rees, M., and Ostriker, J. P.: 1977, *Monthly Notices Roy. Astron. Soc.* **179**, 541–559.
Refsdal, S.: 1970, *Astrophys. J.* **159**, 357–375.
Reinhardt, M.: 1971, *Nature* **229**, 36.
Rindler, W.: 1969, *Essential Relativity*, pp. 12–28, Van Nostrand Reinhold, New York.
Ritter, R. C., and Beams, J. W.: 1978, in *On the Measurement of Cosmological Variations of the Gravitational Constant* (ed. L. Halpern), pp. 29–70, Univ. Presses of Florida, Florida.
Ritter, R. C., Gillies, G. T., Rood, R. T., and Beams, J. W.: 1978, *Nature* **271**, 228–229.
Roach, F. E., and Smith, L. L.: 1968, *Geophys. J. Roy. Astron. Soc.* **15**, 227–239.
Roberts, M. S., and Rots, A. H.: 1973, *Astron. Astrophys.* **26**, 483–485.
Robertson, H. P.: 1933, *Z. Astrophys.* **7**, 153–179.
Robertson, H. P.: 1935, *Astrophys. J.* **82**, 284–301.
Robertson, H. P.: 1936a, *Phys. Rev. (Ser. 2)* **49**, 755–760.
Robertson, H. P.: 1936b, *Astrophys. J.* **83**, 187–201.
Robertson, H. P.: 1936c, *Astrophys. J.* **83**, 257–271.
Roeder, R. C.: 1975, *Astrophys. J.* **196**, 671–673.
Rogstad, D. H., and Shostak, G. S.: 1976, *Astrophys. J.* **176**, 315–321.
Roll, P. G., Krotkov, R., and Dicke, R. H.: 1964, *Ann. Phys. (N.Y.)* **26**, 442–517.
Rood, H. J.: 1965, Thesis, Univ. Michigan.
Rood, H. J.: 1969, *Astrophys. J.* **158**, 657–667.
Rood, H. J.: 1970, *Astrophys. J.* **162**, 333–336.
Rood, H. J.: 1974a, *Astrophys. J.* **188**, 451–461.
Rood, H. J.: 1974b, *Astrophys. J.* **193**, 1–13.
Rood, H. J.: 1974c, *Astrophys. J.* **193**, 15–18.
Rood, H. J.: 1974d, *Astrophys. J.* **194**, 27–35.
Rood, H. J., and Baum, W. A.: 1967, *Astron. J.* **72**, 398–406.
Rood, H. J., and Dickel, J. R.: 1978, *Astrophys. J.* **224**, 724–744.
Rood, H. J., Page, T. L., Kintner, E. C., and King, I. R.: 1972, *Astrophys. J.* **175**, 627–647.
Rood, H. J., Rothman, V. C. A., and Turnrose, B. E.: 1970, *Astrophys. J.* **162**, 411–423.
Rood, H. J., and Sastry, G. N.: 1971, *Pub. Astron. Soc. Pacific* **83**, 313–319.
Rose, M. E.: 1961, *Relativistic Electron Theory*, pp. 117, 119, J. Wiley, New York.
Rosen, G.: 1967, *Can. J. Phys.* **45**, 2383–2384.
Rosen, J., and Rosen, N.: 1975, *Astrophys. J.* **202**, 782–787.
Rosen, J., and Rosen, N.: 1977, *Astrophys. J.* **212**, 605–607.
Rosen, N.: 1940a, *Phys. Rev. (Ser. 2)* **57**, 147–150.
Rosen, N.: 1940b, *Phys. Rev. (Ser. 2)* **57**, 150–153.
Rosen, N.: 1963, *Ann. Phys. (N.Y.)* **22**, 1–11.
Rosen, N.: 1969a, *Nuovo Cimento (Lett., Ser. 1)* **1**, 42–44.
Rosen, N.: 1969b, *Proc. Israel Acad. Sciences and Humanities, Section of Sciences*, No. 12.
Rosen, N.: 1969c, *Int. J. Theor. Phys.* **2**, 189–198.
Rosen, N.: 1971a, *J. Gen. Rel. Grav.* **2**, 129–148.
Rosen, N.: 1971b, *J. Gen. Rel. Grav.* **2**, 223–234.
Rosen, N.: 1973, *J. Gen. Rel. Grav.* **4**, 435–447.
Rosen, N.: 1974, *Ann. Phys. (N.Y.)* **84**, 455–473.
Rosen, N.: 1976, *J. Gen. Rel. Grav.* **7**, 839–840.

Rosen, N.: 1977a, *Astrophys. J.* **211**, 357–360.
Rosen, N.: 1977b, *Nuovo Cimento (Lett., Ser. 2)* **19**, 249–250.
Rosen, N.: 1978, *Astrophys. J.* **221**, 284–285.
Rowan-Robinson, M.: 1972a, *Astron. J.* **77**, 543–549.
Rowan-Robinson, M.: 1972b, *Astrophys. J.* **178**, L81–L83.
Rowan-Robinson, M.: 1977a, *Astrophys. J.* **213**, 635–647.
Rowan-Robinson, M.: 1977b, *Nature* **270**, 9–10.
Roxburgh, I. W.: 1976, *Nature* **261**, 301–302.
Roxburgh, I. W.: 1977a, *Brit. J. Phil. Sci.* **28**, 171–180.
Roxburgh, I. W.: 1977b, *Nature* **265**, 763.
Roxburgh, I. W.: 1977c, *Nature* **268**, 504–507.
Roxburgh, I. W.: 1977d, *Monthly Notices Roy. Astron. Soc.* **181**, 637–645.
Roxburgh, I. W., and Tavakol, R. V.: 1975, *Monthly Notices Roy. Astron. Soc.* **170**, 599–610.
Roxburgh, I. W., and Tavakol, R. K.: 1978, *Found. Phys.* **8**, 229–237.
Ruban, V. A.: 1977, *Astrophys. Space Sci.* **46**, L23–L28.
Rubin, V. C., Ford, W. K., and Rubin, J. S.: 1973, *Astrophys. J.* **183**, L111–L115.
Rubin, V. C., Ford, W. K., Thonnard, N., Roberts, M. S., and Graham, J. A.: 1976a, *Astron. J.* **81**, 687–718.
Rubin, V. C., Thonnard, N., Ford, W. K., and Roberts, M. S.: 1976b, *Astron. J.* **81**, 719–737.
Runcorn, S. K.: 1955, *Endeavor* **14**, 152–159.
Runcorn, S. K.: 1964, *Nature* **204**, 823–825.
Runcorn, S. K. (ed.): 1967, *Mantles of the Earth and Terrestrial Planets*, Interscience Publ., London.
Russel, C. T.: 1978, *Nature* **272**, 147–148.
Sadeh, D., Knowles, S. H., and Yaplee, B. S.: 1968a, *Science* **159**, 307–308.
Sadeh, D., Knowles, S. H., and Au, B.: 1968b, *Science* **161**, 567–569.
Sadeh, D., Hollinger, J. P., Knowles, S. H., and Youmans, A. B.: 1968c, *Science* **162**, 897–898.
Safranov, V. S.: 1978, *Icarus* **33**, 3–12.
Sagan, C.: 1977, *Nature* **269**, 224–226.
Sagan, C., and Mullen, G.: 1972, *Science* **177**, 52–56.
Sakharov, A. D.: 1968, *Soviet Phys Dokl.* **12**, 1040–1041.
Salam, A.: 1968, in *Elementary Particle Theory* (ed. N. Svartholm), pp. 367–377, Almqvist and Wiksell, Stockholm.
Salam, A.: 1972, in *Developments in High Energy Physics* (ed. R. Gatto), pp. 386–426, Academic Press, New York.
Salmona, A.: 1967, *Phys. Rev. (Ser. 2)* **154**, 1218–1223.
Sandage, A., Kristian, J., and Westphal, J. A.: 1976, *Astrophys. J.* **205**, 688–695.
Sandage, A., and Tamman, G.: 1974, *Astrophys. J.* **194**, 559–568.
Sandage, A., and Tamman, G.: 1975a, *Astrophys. J.* **196**, 313–328.
Sandage, A., and Tamman, G.: 1975b, *Astrophys. J.* **197**, 265–280.
Saslaw, W. C.: 1977, *Monthly Notices Roy. Astron. Soc.* **179**, 659–662.
Schatten, K. H.: 1975, *Astrophys. Space Sci.* **34**, 467–480.
Schatten, K. H.: 1977, *Astrophys. J.* **216**, 650–653.
Schiff, L. I.: 1939, *Proc. Nat. Acad. Sci. (U.S.A.)* **25**, 391–395.
Schiff, L. I.: 1960, *Am. J. Phys.* **28**, 340–343.
Schmidt, M.: 1968, *Astrophys. J.* **151**, 393–409.
Schmidt, M.: 1970, *Astrophys. J.* **162**, 371–379.
Schmidt, M.: 1974, in *Astrophysics and Gravitation* (Proc. 16th. Solvay Conf. on Physics), pp. 463–471, Université de Bruxelles, Belgium.

Schmidt, M.: 1975, in *Galaxies and the Universe* (Vol. 9, *Stars and Stellar Systems*, eds. A. Sandage, M. Sandage, and J. Kristian), pp. 283–308, Univ. Chicago Press, Chicago, Ill.
Sciama, D. W.: 1967, *Phys. Rev. Lett.* **18**, 1065–1067.
Sciama, D. W.: 1978, *New Scientist* **77**, 298–300.
Sears, D. W.: 1978, *The Nature and Origin of Meteorites*, A. Hilger, Bristol, U.K.
Sedov, L. I.: 1959, *Similarity and Dimensional Methods in Mechanics*, Academic Press, London.
Segal, I. E.: 1972, *Astron. Astrophys.* **18**, 143–148.
Segal, I. E.: 1974, *Proc. Nat. Acad. Sci. (U.S.A.)* **71**, 765–768.
Segal, I. E.: 1975, *Proc. Nat. Acad. Sci. (U.S.A.)* **72**, 2473–2477.
Segal, I. E.: 1976a, *Mathematical Cosmology and Extragalactic Astronomy*, Academic Press, New York.
Segal, I. E.: 1976b, *Proc. Nat. Acad. Sci. (U.S.A.)* **73**, 669–673.
Segal, I. E.: 1976c, *Proc. Nat. Acad. Sci. (U.S.A.)* **73**, 3355–3359.
Segal, I. E.: 1977, in *Proc. Fifth International Colloquium on Group Theoretical Methods in Physics* (Montreal, July 1976; eds. R. T. Sharp and B. Kolman), pp. 433–447, Academic Press, New York.
Segal, I. E.: 1978a, *Astron. Astrophys.* **68**, 343–344.
Segal, I. E.: 1978b, *Astron. Astrophys.* **68**, 353–359.
Selak, S.: 1978, *Astrophys. Space Sci.* **56**, 275–284.
Seldner, M.: 1977, Dissertation, Princeton Univ.
Seldner, M., and Peebles, P. J. E.: 1977, *Astrophys. J.* **215**, 703–716.
Seldner, M., and Peebles, P. J. E.: 1978, *Astrophys. J.* **225**, 7–20.
Seldner, M., Siebers, B., Groth, E. J., and Peebles, P. J. E.: 1977, *Astron. J.* **82**, 249–256.
Setti, G., and Woltjer, L.: 1973, *Astrophys J.* **181**, L61–L63.
Setti, G., and Woltjer, L.: 1977, *Astrophys J.* **218**, L33–L35.
Sexl, R.: 1972, in *Developments in High Energy Physics* (ed. R. Gatto), pp. 331–371, Academic Press, New York.
Shäfer, G., and Dehnen, H.: 1977a, *Astron. Astrophys.* **54**, 823–836.
Shäfer, G., and Dehnen, H.: 1977b, *Astron. Astrophys.* **61**, 671–677.
Shakeshaft, J. R. (ed.): 1974, *The Formation and Dynamics of Galaxies* (I.A.U. Symp. 58), D. Reidel, Dordrecht, Holland.
Shamir, I.: 1969, *Nature* **222**, 362.
Shamir, J., and Fox, R.: 1967, *Nuovo Cimento (Ser. 10)* **B50**, 371–372.
Shane, C. D.: 1975, in *Galaxies and the Universe* (Vol. 9, *Stars and Stellar Systems*, eds. A. Sandage, M. Sandage, and J. Kristian), pp. 647–663, Univ. Chicago Press, Chicago, Ill.
Shapiro, I., Pettengill, G. H., Ash, M. E., Stone, M. L., Smith, W. B., Ingalls, R. P., and Brockelman, R. A.: 1968, *Phys. Rev. Lett.* **20**, 1265–1269.
Shapiro, I., Counselman, C. C., and King, R. W.: 1976, *Phys. Rev. Lett.* **36**, 555–558.
Shapiro, I., Smith, W. B., Ash, M. B., Ingalls, R. P., and Pettengill, G. H.: 1971, *Phys. Rev. Lett.* **26**, 27–30.
Shapiro, S. L.: 1971, *Astron. J.* **76**, 291–293.
Shaw, G. H.: 1978, *Phys. Earth Planet Interiors* **16**, 361–369.
Shectman, S. A.: 1973, *Astrophys. J.* **179**, 681–698.
Shectman, S. A.: 1974, *Astrophys. J.* **188**, 233–242.
Shepley, L. C.: 1973, in *Cargèse Lectures in Physics* 6 (ed. E. Schatzman), pp. 227–260, Gordon and Breach, London.
Shields, G. A.: 1978, *Nature* **273**, 519–520.
Shouten, J. A.: 1949, *Rev. Mod. Phys.* **21**, 421–424.
Shouten, J. A., and Haantjes, J.: 1936, *Proc. Kon. Akad. Wet.* **39**, 1059–1065.
Shouten, J. A., and Van Dantzig, D.: 1932, *Z. Phys.* **78**, 639–667.

Silk, J.: 1968, *Astrophys. J.* **151**, 459–471.
Silk, J.: 1970, *Monthly Notices Roy. Astron. Soc.* **147**, 13–19.
Silk, J.: 1973, in *Cargèse Lectures in Physics* **6** (ed. E. Schatzman), pp. 401–503, Gordon and Breach, London.
Silk, J.: 1974a, *Astrophys. J.* **193**, 525–527.
Silk, J.: 1974b, Gravity Research Foundation 1st. Award Essay.
Silk, J.: 1977a, *Astrophys. J.* **211**, 638–648.
Silk, J.: 1977b, *Astrophys. J.* **214**, 152–160.
Silk, J.: 1977c, *Astrophys. J.* **214**, 718–724.
Silk, J.: 1977d, *Nature* **265**, 710–711.
Silk, J.: 1977e, *Astron. Astrophys.* **59**, 53–58.
Silk, J., and Lea, S.: 1973, *Astrophys. J.* **180**, 669–686.
Silk, J., and White, S. D. M.: 1978, *Astrophys. J.* **223**, L59–L62.
Silk, S., and Ames, S.: 1972, *Astrophys. J.* **178**, 77–93.
Sill, G. T., and Wilkening, L. L.: 1978, *Icarus* **33**, 13–22.
Simons, S.: 1976a, *Astrophys. Space Sci.* **41**, 423–434.
Simons, S.: 1976b, *Astrophys. Space Sci.* **41**, 435–445.
Simpson, I. C.: 1978, *Astrophys. Space Sci.* **57**, 381–400.
Sirag, S.-P.: 1977, *Nature* **268**, 294.
Slattery, W. L.: 1978, *Moon and Planets* **19**, 443–456.
Slattery, W. L., Cameron, A. G. W., and DeCampli, W. M.: 1978, *Bull. Am. Astron. Soc.* **10**, 591.
Smalley, L. L., and Eby, P. B.: 1976, *Nuovo Cimento (Ser. 11)* **B35**, 54–60.
Smalley, L. L.: 1978, *Found Phys.* **8**, 59–68.
Smith, H. M.: 1969, *Nature* **221**, 221–223.
Smith, P. J.: 1978, *Nature* **271**, 301.
Smoot, G. F., Gorenstein, M. V., and Muller, R. A.: 1977, *Phys. Rev. Lett.* **39**, 898–901.
Soneira, R. M., and Peebles, P. J. E.: 1977, *Astrophys. J.* **211**, 1–15.
Soneira, R. M., and Peebles, P. J. E.: 1978, *Astron. J.* **83**, 845–861.
Spinrad, H., Ostriker, J., Stone, R. P. S., Chiu, L.-T. G., and Bruzual, G.: 1978, *Astrophys. J.* **225**, 56–66.
Spinrad, H., and Stone, R. P. S.: 1978, *Astrophys. J.* **226**, 609–612.
Stacey, F. D.: 1978, *Geophys. Res. Lett.* **5**, 377–378.
Stannard, D.: 1973, *Nature* **246**, 295–297.
Stecker, F. W.: 1978, *Nature* **273**, 493–497.
Steeruwitz, E.: 1975, *Phys. Rev. (Ser. 3)* **D11**, 3378–3383.
Steigman, G.: 1974, in *Confrontation of Cosmological Theories with Observational Data* (I.A.U. Symp. 63, ed. M. Longair), pp. 347–356, D. Reidel, Dordrecht, Holland.
Steigman, G.: 1978, *Astrophys. J.* **221**, 407–411.
Steiner, J.: 1977, *Geology* **5**, 313–318.
Stephenson, C. B.: 1977, *Astrophys. Space Sci.* **51**, 117–119.
Stephenson, F. R., and Clark, D. H.: 1978, *Applications of Early Astronomical Records*, A. Hilger, Bristol, U.K.
Stephenson, L. and Cohn, J.: 1978a, *J. Gen. Rel. Grav.* **9**, 21–38.
Stephenson, L. and Cohn, J.: 1978b, *J. Gen. Rel. Grav.* **9**, 39–51.
Stewart, A. D.: 1977, *J. Geol. Soc. London* **133**, 281–291.
Stewart, A. D.: 1978, *Nature* **271**, 153–155.
Stewart, J. M., and Sciama, D. W.: 1967, *Nature* **216**, 748–753.
Stewart, N. J., and Hawkins, M. R. S.: 1978, *Nature* **276**, 163–165.
Stockton, A.: 1978, *Astrophys. J.* **223**, 747–757.
Stoeger, W. R.: 1978, *J. Gen. Rel. Grav.* **9**, 165–174.

Stolz, A., Bender, P. L., Faller, J. E., Silverberg, E. C., Mulholland, J. D., Shelus, P. J., Williams, J. G., Carter, W. E., Currie, D. G., and Kaula, W. M.: 1976, *Science* **193**, 997–999.
Strangway, D. W.: 1978, *Moon and Planets* **18**, 273–279.
Strnad, J.: 1970, *Nature* **226**, 137–138.
Sunyaev, R. A., and Zel'dovich, Y. B.: 1972, *Comm. Astrophys. Space Phys.* **4**, 173–178.
Surdin, M.: 1978, *Found. Phys.* **8**, 341–357.
Swann, W. F. G.: 1927, *Phil. Mag. (Ser. 7)*, **3**, 1088–1136.
Synge, J. L.: 1966, *Relativity: The General Theory*, pp. 317–322, North-Holland Publ. Co. Amsterdam.
Synge, J. L.: 1969, *Nature* **223**, 161–162.
Szekeres, G.: 1955, *Phys. Rev. (Ser. 2)* **97**, 212–223.
Szekeres, G.: 1957, *J. Math. Mech.* **6**, 471–517.
Szekeres, G.: 1968, *Nature* **220**, 1116–1118.
Szekeres, P.: 1975, *Comm. Math. Phys.* **41**, 55–64.
Taylor, B. N., Parker, W. H., and Langenberg, D. N.: 1969, *The Fundamental Constants and Quantum Electrodynamics*, Academic Press, New York.
Taylor, J.: 1976, *Gauge Theories of the Weak Interactions*, Cambridge Univ. Press, Cambridge.
Taylor, N. W.: 1962, *Nature* **194**, 647–649.
Teerikorpi, P.: 1978, *Astron. Astrophys.* **64**, 379–388.
Teerikorpi, P., and Jaakkola, T.: 1977, *Astron. Astrophys.* **59**, L33–L36.
Teller, E.: 1948, *Phys. Rev. (Ser. 2)* **73**, 801–802.
Ter Haar, D.: 1954a, *Phil. Mag. (Ser. 7)* **45**, 320–324.
Ter Haar, D.: 1954b, *Phil. Mag. (Ser. 7)* **45**, 1023–1024.
Thirring, H.: 1918, *Phys. Z.* **19**, 33–39.
Thirring, H.: 1921, *Phys. Z.* **22**, 29–30.
Thorne, K. S., and Dykla, J. J.: 1971, *Astrophys. J.* **166**, L35–L38.
Thuan, T. X., Hart, M. H., and Ostriker, J.: 1975, *Astrophys. J.* **201**, 756–772.
Tifft, W. G.: 1973, *Astrophys. J.* **181**, 305–326.
Tifft, W. G.: 1974, in *The Formation and Dynamics of Galaxies* (I.A.U. Symp. 58, ed. J. R. Shakeshaft), pp. 243–256, D. Reidel, Dordrecht, Holland.
Tifft, W. G.: 1976, *Astrophys. J.* **206**, 38–56.
Tifft, W. G.: 1977, *Astrophys. J.* **211**, 31–46.
Tifft, W. G.: 1978, *Astrophys. J.* **220**, 418–425.
Tifft, W. G., and Gregory, S. A.: 1976, *Astrophys. J.* **205**, 696–708.
Tinsley, B.: 1977a, *Phys. Today* **30** (June), 32–38.
Tinsley, B.: 1977b, *Astrophys. J.* **211**, 621–637 (erratum, ibid, **216**, 349–350, 1977).
Tinsley, B.: 1978a, *Nature* **273**, 208–211.
Tinsley, B.: 1978b, in *The Large Scale Structure of the Universe* (I.A.U. Symp. 79, eds. M. Longair and J. Einasto), pp. 343–355, D. Reidel, Dordrecht, Holland.
Tinsley, B.: 1978c, *Astrophys. J.* **220**, 816–821.
Tipler, F. J.: 1976, *Astrophys. J.* **209**, 12–15.
Toksöz, M. N., and Hsui, A. T.: 1978, *Icarus* **34**, 537–547.
Toksöz, M. N., Hsui, A. T., and Johnston, D. H.: 1978, *Moon and Planets* **18**, 281–320.
Tomimura, N.: 1977, *Nuovo Cimento (Ser. 11)* **B42**, 1–8.
Tomita, K.: 1972, *Prog. Theor. Phys.* **48**, 1503–1516.
Tomita, K.: 1973, *Prog. Theor. Phys.* **50**, 1285–1301.
Tomita, K.: 1977, in *Cosmology, History and Theology* (eds. W. Yourgrau and A. D. Breck), pp. 131–139, Plenum Press, New York.
Trautman, A.: 1973, *Nature Phys. Sci.* **242**, 7–8.

Tully, R. B., and Fisher, J. R.: 1978, in *The Large Scale Structure of the Universe* (I.A.U. Symp. 79, eds. M. Longair and J. Einasto), pp. 31–47, D. Reidel, Dordrecht, Holland.
Turner, E. L.: 1976a, *Astrophys. J.* **208**, 20–29.
Turner, E. L.: 1976b, *Astrophys. J.* **208**, 304–316.
Turner, E. L., and Gott, J. R.: 1975, *Astrophys. J.* **197**, L89–L93.
Turner, E. L., and Gott, J. R.: 1976a, *Astrophys. J. Suppl.* **32**, 409–427.
Turner, E. L., and Gott, J. R.: 1976b, *Astrophys. J.* **209**, 6–11.
Unruh, W. G.: 1976, *Phys. Rev. (Ser. 3)* **D14**, 870–892.
Van Dantzig, D.: 1932, *Math. Ann.* **106**, 400–454.
VandenBerg, D. A.: 1976, *Monthly Notices Roy. Astron. Soc.* **176**, 455–461.
VandenBerg, D. A.: 1977, *Monthly Notices Roy. Astron. Soc.* **181**, 695–701.
Van Diggelen, J.: 1976, *Nature* **262**, 675–676.
Van Flandern, T. C.: 1970, *Astron. J.* **75**, 657–658.
Van Flandern, T. C.: 1975, *Monthly Notices Roy. Astron. Soc.* **170**, 333–342.
Van Flandern, T. C.: 1976, *Sci. Am.* **234** (Feb.), 44–52.
Van Flandern, T. C.: 1978, in *On the Measurement of Cosmological Variations of the Gravitational Constant* (ed. L. Halpern), pp. 21–28, Univ. Presses of Florida, Florida.
Van Hilten, D.: 1965, *Geophys. J. Roy. Astron. Soc.* **9**, 279–281.
Van Hilten, D.: 1968, *Tectonophysics* **5**, 191–210.
Varshni, Y. P.: 1975, *Astrophys. Space Sci.* **37**, L1–L6.
Varshni, Y. P.: 1976, *Astrophys. Space Sci.* **43**, 3–8.
Varshni, Y. P.: 1977, *Astrophys. Space Sci.* **51**, 121–124.
Veblen, O.: 1929, *J. London Math. Soc.* **4**, 140–160.
Veblen, O.: 1930, *Q. J. Maths. (Oxford Ser.)* **1**, 60–76.
Veblen, O., and Hoffmann, B.: 1930, *Phys. Rev. (Ser. 2)* **36**, 810–822.
Verhulst, F.: 1975, *Celes. Mech.* **11**, 95–129.
Vigier, J.-P.: 1977, in *Cosmology, History and Theology* (eds. W. Yourgrau and A. D. Breck), pp. 141–157, Plenum Press, New York.
Vila, S. C.: 1976, *Astrophys. J.* **206**, 213–214.
Vinti, J. P.: 1974, *Monthly Notices Roy. Astron. Soc.* **169**, 417–427.
Vollmer, R.: 1977, *Nature* **270**, 144–147.
Von Hoerner, S.: 1973a, *Astrophys. J.* **186**, 741–765.
Von Hoerner, S.: 1973b, *Astrophys. J.* **181**, 261–265.
Wagoner, R. V.: 1970, *Phys. Rev. (Ser. 3)* **D1**, 3209–3216.
Wald, R. M.: 1978, *Ann. Phys. (N.Y.)* **110**, 472–486.
Walker, A. G.: 1935, *Monthly Notices Roy. Astron. Soc.* **95**, 263–269.
Walker, A. G.: 1936, *Proc. London Math. Soc.* **42**, 90–127.
Walker, A. G.: 1941, *Observatory* **64**, 17–23.
Walker, A. G.: 1945, *Proc. Roy. Soc. Edinburgh* **A62**, 164–174.
Wampler, E. J., Baldwin, J. A., Burke, W. L., Robinson, L. B., and Hazard, C.: 1973, *Nature* **246**, 203–205.
Wapstra, A. H., and Nijgh, G. J.: 1955, *Physica* **21**, 796–798.
Ward, A.: 1961, *Nature* **192**, 858.
Ward, M. A.: 1963, *Geophys. J. Roy. Astron. Soc.* **8**, 217–225.
Ward, M. A.: 1966, *Geophys. J. Roy. Astron. Soc.* **10**, 445–447.
Ward, W. R.: 1976, in *Frontiers of Astrophysics* (ed. E. H. Avrett), pp. 1–40, Harvard Univ. Press, Cambridge, Mass.
Warner, B., and Nather, R. E.: 1969, *Nature* **222**, 157–158.
Wasserman, I.: 1978, *Astrophys. J.* **224**, 337–343.
Wasserman, I., and Brecher, K.: 1978, *Phys. Rev. Lett.* **41**, 920–923.
Wdowczyk, J., and Wolfendale, A. W.: 1975, *Nature* **258**, 217–218.
Weidenschilling, S. J.: 1978, *Icarus* **35**, 99–111.

Weinberg, S.: 1967, *Phys. Rev. Lett.* **19**, 1264–1266.
Weinberg, S.: 1972, *Gravitation and Cosmology*, J. Wiley, New York.
Weinberg, S.: 1974a, *Rev. Mod. Phys.* **46**, 255–277.
Weinberg, S.: 1974b, *Phys. Rev. (Ser. 3)* **D9**, 3357–3378.
Weinberg, S.: 1976a, *Phys. Rev. Lett.* **36**, 294–296.
Weinberg, S.: 1976b, *Astrophys. J.* **208**, L1–L3.
Weinberg, S.: 1977a, *Phys. Today* **30** (April), 42–50.
Weinberg, S.: 1977b, *Am. Sci.* **65** (March/April), 171–176.
Weinberg, S.: 1977c, *Daedelus* **2**, 17–35.
Weinberg, S.: 1977d, *The First Three Minutes*, A. Deutsch, London.
Weinberg, S.: 1979, *Phys. Rev. Lett.* **42**, 850–853.
Weinstein, D. H., and Keeney, J.: 1973a, *Nuovo Cimento (Lett., Ser. 2)* **8**, 299–302.
Weinstein, D. H., and Keeney, J.: 1973b, *Nature* **244**, 83–84.
Weinstein, D. H., and Keeney, J.: 1974, *Nature* **247**, 140.
Weinstein, D. H., and Keeney, J.: 1975, in *Growth Rhythms and the History of the Earth's Rotation* (eds. G. D. Rosenberg and S. K. Runcorn), pp. 377–384, J. Wiley, New York.
Wesson, P. S.: 1973, *Q. J. Roy. Astron. Soc.* **14**, 9–64.
Wesson, P. S.: 1975a, in *Growth Rhythms and the History of the Earth's Rotation* (eds. G. D. Rosenberg and S. K. Runcorn), pp. 353–374, J. Wiley, New York.
Wesson, P. S.: 1975b, *Astrophys. Space Sci.* **36**, 363–382.
Wesson, P. S.: 1976, *Astrophys. Space Sci.* **40**, 325–349.
Wesson, P. S.: 1977a, *Astronomisk Tidsskrift* **10**, 162–163.
Wesson, P. S.: 1977b, *Astron. Astrophys.* **61**, 177–180.
Wesson, P. S.: 1978a, *Cosmology and Geophysics*, A. Hilger, Bristol, U.K.
Wesson, P. S.: 1978b, *Nature* **273**, 572.
Wesson, P. S.: 1978c, *Astron. Astrophys.* **68**, 131–137.
Wesson, P. S.: 1978d, *Astrophys. Space Sci.* **54**, 489–495.
Wesson, P. S.: 1978e, *Astrophys. Lett.* **19**, 121–125.
Wesson, P. S.: 1978f, *Astron. Astrophys.* **69**, 125–128.
Wesson, P. S.: 1978g, *Astrophys. Lett.* **19**, 127–131.
Wesson, P. S.: 1978h, *J. Math. Phys.* **19**, 2283–2284.
Wesson, P. S.: 1979a, *Astrophys. J.* **228**, 647–663.
Wesson, P. S.: 1979b, Thesis, Cambridge Univ.
Wesson, P. S.: 1979c, *Astron. Astrophys.* **80**, 296–300.
Wesson, P. S.: 1979d, *Astron. Astrophys.* **76**, 200–207.
Wesson, P. S., and Lermann, A.: 1976, *Astron. Astrophys.* **53**, 383–388.
Wesson, P. S., and Lermann, A.: 1977a, *Astrophys. Space Sci.* **46**, 327–334.
Wesson, P. S., and Lermann, A.: 1977b, *Astrophys. Space Sci.* **46**, 51–60.
Wesson, P. S., and Lermann, A.: 1978a, *Astrophys. Space Sci.* **57**, 203–217 (erratum, ibid, **60**, 242, 1979).
Wesson, P. S., and Lermann, A.: 1978b, *Icarus* **33**, 74–88.
Wesson, P. S., Lermann, A., and Goodson, R. E.: 1977, *Astrophys. Space Sci.* **48**, 357–363.
Weyl, H.: 1919, *Ann. Phys. (Ser. 4)* **59**, 101–133.
Weyl, H.: 1922, *Space-Time-Matter*, Methuen, London.
Weymann, R. J., Boroson, T., and Scargle, J. D.: 1978, *Astrophys. Space Sci.* **53**, 265–278.
Wheelon, A. D.: 1952, *Phys. Rev. (Ser. 2)* **85**, 383–384.
Wheeler, J. A.: 1962, *Geometrodynamics*, Academic Press, New York.
White, S. D. M.: 1977, *Comm. Astrophys.* **7**, 95–102.
White, S. D. M., and Rees, M.: 1978, *Monthly Notices Roy. Astron. Soc.* **183**, 341–358.
White, S. D. M., and Sharp, N. A.: 1977, *Nature* **269**, 395–396.
Whitrow, G. J., and Yallop, B. D.: 1964, *Monthly Notices Roy. Astron. Soc.* **127**, 301–318.
Whitrow, G. J., and Yallop, B. D.: 1965, *Monthly Notices Roy. Astron. Soc.* **130**, 31–43.

Wickramasinghe, N. C., Edmunds, M. G., Chitre, S. M., Narlikar, J. V., and Ramadurai, S.: 1975, *Astrophys. Space Sci.* **35**, L9–L13.
Wilkie, T.: 1977, *Nature* **268**, 295–296.
Wilkinson, D. H.: 1975, *Nature* **257**, 189–193.
Will, C. M.: 1971, *Astrophys. J.* **169**, 125–140.
Will, C. M.: 1974, *Phys. Rev. (Ser. 3)* **D10**, 2330–2337.
Will, C. M., and Eardley, D. M.: 1977, *Astrophys. J.* **212**, L91–L94.
Williams, E., Faller, J., and Hill, H.: 1971, *Phys. Rev. Lett.* **26**, 721–724.
Williams, J. G., Dicke, R. H., Bender, P. L., Alley, C. O., Carter, W. E., Currie, D. G., Eckhardt, D. H., Faller, J. E., Kaula, W. M., Mulholland, J. D., Plotkin, H. H., Poultney, S. K., Shelus, P. J., Silverberg, E. C., Sinclair, W. S., Slade, M. A., and Wilkinson, D. T.: 1976, *Phys. Rev. Lett.* **36**, 551–554.
Wills, D.: 1978, *Physica Scripta* **17**, 333–337.
Wills, D., and Lynds, R.: 1978, *Astrophys. J. Suppl.* **36**, 317–358.
Wills, D., and Ricklefs, R. L.: 1976, *Monthly Notices Roy. Astron. Soc.* **175**, 81P–85P.
Windley, B. F. (ed.): 1976, *The Early History of the Earth*, J. Wiley, London.
Winkler, G. M. R., and Van Flandern, T. C.: 1977, *Astron. J.* **82**, 84–92.
Witten, L.: 1962, in *Gravitation: An Introduction to Current Research* (ed. L. Witten), pp. 382–411, J. Wiley, New York.
Woodward, J. F., and Crowley, R. J.: 1973, *Nature Phys. Sci.* **246**, 41.
Woodward, J. F., and Yourgrau, W.: 1970a, *Nature* **226**, 619–621.
Woodward, J. F., and Yourgrau, W.: 1970b, *Ann. Phys. (Ser. 7)* **25**, 334–336.
Woodward, J. F., and Yourgrau, W.: 1971, *Nature* **229**, 36–37.
Woodward, J. F., and Yourgrau, W.: 1972, *Nuovo Cimento (Ser. 11)* **B9**, 440–452.
Woodward, J. F., and Yourgrau, W.: 1973, *Nature* **241**, 338.
Woolfson, M.: 1969, *Rep. Prog. Phys.* **32**, 135–185.
Wrigley, W. (ed.): 1978, 'Selected Papers from the 2nd. and 3rd. Space Relativity Symposia, Lisbon (1975) and Anaheim, California (1976)'. *Acta Astronautica* **5**(1/2), 3–130.
Wyler, A.: 1969, *Acad. Sci. Paris, Comptes Rendus* **A269**, 743–745.
Wyler, A.: 1971, *Acad. Sci. Paris, Comptes Rendus* **A271**, 186.
Wyman, M.: 1946, *Phys. Rev. (Ser. 2)* **70**, 396–400.
Wyman, M.: 1976, *Can. Math. Bull.* **19**, 343–357.
Wyman, M.: 1978, *Aust. J. Phys.* **31**, 111–114.
Yabushita, S.: 1964, *Nuovo Cimento (Ser. 10)* **33**, 769–775.
Yahil, A.: 1972, *Astrophys. J.* **178**, 45–55.
Yahil, A.: 1974, in *The Interaction Between Science and Philosophy* (ed. Y. Elkana), pp. 27–33, Humanities Press, Atlantic Highlands, N.J.
Yang, K.-H.: 1976a, *Ann. Phys. (N.Y.)* **101**, 62–96.
Yang, K.-H.: 1976b, *Ann. Phys. (N.Y.)* **101**, 97–118.
Yilmaz, Y.: 1975, *J. Gen. Rel. Grav.* **6**, 269–276.
Yokoi, K.: 1972, *Prog. Theor. Phys.* **48**, 1760–1761.
Yoshimura, M.: 1978, *Phys. Rev. Lett.* **41**, 281–284.
Zel'dovich, Y. B.: 1963, *Soviet Phys. JETP* **16**, 1395–1396.
Zel'dovich, Y. B.: 1964, *Soviet Astron.* **8**, 13–16.
Zel'dovich, Y. B., and Novikov, I. D.: 1971, *Relativistic Astrophysics*, Vol. 1, Chicago Univ. Press, Chicago, Ill.
Zel'dovich, Y. B., and Novikov, I. D.: 1975, *Structure and Evolution of the Universe*, Nauka, Moscow.
Zel'manov, A. L.: 1977, *Soviet Astron.* **21**, 664–671.
Zwicky, F.: 1961, *Pub. Astron. Soc. Pacific* **73**, 314–317.
Zwicky, F.: 1971, *Catalogue of Selected Compact Galaxies and of Post-Eruptive Galaxies*, Zurich, Switzerland.

INDEX

Absorber theory 19–20, 23–24, 108
Angular momentum 25, 63–64

Bergmann/Wagoner theory 91–92
Bicknell/Klotz theory 92
Black holes 35, 39, 67, 90, 105, 126, 130, 132
Brans/Dicke theory 5, 10, 22, 59, 89–92, 104, 106–107, 112

Canuto et al theory 24–28, 28–30, 49
Clube theory 100
Clusters, galaxies 13, 41, 71, 82, 101, 109, 144, 147–152
Clusters, stars 41–44
Conformal group 96, 114–121
Conformal invariance 19, 28, 92, 106, 115–118
Continental drift 50, 52
Continuous creation 11–19, 20–21, 25–28, 42–43, 48–52, 55, 61–62, 64, 107–109
Convection currents 52, 76–77
Cosmic rays 133–134, 153
Cosmological coincidences 131–134
Cosmological constant 6, 10–11, 26, 32–37, 89, 92, 116, 151
Cosmological Principle 67
Cosmology, baryon-symmetric 109–110
Cosmology, homogeneous 26, 68, 104–105, 107–108, 111, 121, 150, 156
Cosmology, inhomogeneous 67–70, 143, 147, 150–151, 158, 161
Cosmology, phase changes 26, 33–34, 37–39
Cosmology, scale-free 26, 66–70

Davies/Unruh effect 36
Day, length of 25
Dirac theory 9–19, 28–30, 47, 61, 64, 80–83, 107, 122–123, 127

Earth, expansion 21–22, 25, 48–56, 60, 63–64, 72, 81, 84
Earth, evolution 52–55, 74–79
Earth, phase changes 53–54

Earth-Moon system 48–49, 53–54, 62–66, 71–72, 142
Earth, rotation 25, 48–49, 53–54, 63–66
Earth, surface temperature 12, 14–15, 21–23, 26–27, 42, 45, 77–78
Eddington numbers 123, 131–134
Einstein's equations 6
Electron charge 9, 42, 44–46
Electromagnetism 19–20, 23–24, 26, 32, 37–38, 83, 106–108, 114–121, 130, 139–142
Elementary particles 26, 31–40

Fine-structure constant 42, 46, 123–124
Freund et al theory 100–102

$\dot{G}$, space variability 102
$\dot{G}$, experiments 58–66
$\dot{G}$, limits on 29, 41–44, 48, 54, 57–59, 62, 64
$\dot{G}$, possible value of 29, 48–49, 62–63, 65, 95, 105
$\dot{G}$, predicted value of 9, 22, 25, 28–29, 57
Galaxies, brightness 12, 21, 42–43
Galaxies, formation 38–39, 67, 132, 150–152
Gauge invariance, gravity 10–11, 24–28, 70–73, 80, 83, 106–107, 124–125
Gauge invariance, particles 31–32, 39–40, 83
Gauge theories, gravity 9–30, 70–73, 80, 106–107
Gauge theories, particles 26, 31–40, 109–110
General relativity 5–8, 59, 61, 67, 80, 82, 89–92, 93, 104, 107–110, 119, 122, 129
Goded theory 112
Group theory 96, 111, 114–134

Hoyle/Narlikar theory 19–24, 28–30, 49, 129, 139, 144
Hubble's law 12–13, 68, 71–72, 82, 98, 112, 152–161

Inertia 37

INDEX

Jordan/Brans/Dicke theory 91

Katz theory 107–111
Kinematic relativity 8, 114

Large Numbers Hypothesis 9, 15–18, 24–27, 29–30, 42, 44–46, 49, 60, 64, 82–83, 122–123, 131
Life 22–23, 132–133

Mach's Principle 6, 27–28
Malin theory 111–112
Mass, variation of 11–12, 19–21, 24–27, 41–42, 61–62, 83, 110, 111–112
Meteorites 42, 74, 76
Metric 5, 71, 103
Microwave background 15–18, 21, 27, 34–35, 68, 108, 133–134, 136–138, 153–161
Missing mass problem 13, 40, 44, 95, 101, 128, 139

Neutron stars 99, 104
Ni theory 98–99
Non-Doppler redshifts 20, 96–98, 121, 135–161
Nordtvedt theory 91–92
Nucleosynthesis 33, 38, 40, 44

Omote theory 106–107

Palaeomagnetism 49–52
Palaeontological data 48, 63–64
Photon/baryon ratio 35, 110, 132
Photon, mass 139–143
Planck's constant 23–24, 46
Plate tectonics 76–77
Principle of Equivalence 6–8, 103–104, 112, 122
Prokhovnik theory 13, 93–96
Pulsars 18, 105–106

QSOs 16, 20, 67–70, 98, 111, 142, 143–146, 149

Quantization, gravity 82, 101, 127–131

Rastall theory 99–100
Rosen theory 102–106
Rubin/Ford/Rubin anomaly 152–161

Salam/Weinberg theory 32–40
Scalar-tensor theories 5, 59, 89–92
Scale invariance 10–11, 24–28, 66–80, 89–92, 106–107
Segal theory 96–98, 114, 121
Self-similarity 26, 66–80
Solar System, G effects 25, 29, 58–60, 90–92
Solar System, origin 73–79
Stars, luminosity 11–12, 13, 21, 41–42
Steady state model 17, 107–108
Strong gravity 124–128
Strong interaction 32, 38–39, 83
Sun, evolution 11–12, 18, 22, 25, 41, 45, 77–78
Superclusters 98, 101, 147–148, 150, 154–161
Symmetry breaking 10–11, 33–35, 37–38, 128

Tired-light theory 135–143

Universe, hierarchical 147–149, 151–152, 161
Universe, homogeneous 84, 107–108, 149–150, 159–160
Universe, inhomogeneous 39, 69–70, 82, 84, 147–152, 158, 160–161

Vacuum elasticity 125–126
Vector-tensor theory 5

Weak interaction 32, 37–40, 83, 126, 130
Weinberg/Salam theory 32–40
Weyl geometry 9–14, 24, 83, 106–107, 116, 119, 124–125, 127
White dwarfs 18, 43, 102

ASTROPHYSICS AND SPACE SCIENCE LIBRARY

Edited by

J. E. Blamont, R. L. F. Boyd, L. Goldberg, C. de Jager, Z. Kopal, G. H. Ludwig, R. Lüst,
B. M. McCormac, H. E. Newell, L. I. Sedov, Z. Švestka, and W. de Graaff

1. C. de Jager (ed.), *The Solar Spectrum, Proceedings of the Symposium held at the University of Utrecht, 26–31 August, 1963.* 1965, XIV + 417 pp.
2. J. Orthner and H. Maseland (eds.), *Introduction to Solar Terrestrial Relations, Proceedings of the Summer School in Space Physics held in Alpbach, Austria, July 15–August 10, 1963 and Organized by the European Preparatory Commission for Space Research.* 1965, IX + 506 pp.
3. C. C. Chang and S. S. Huang (eds.), *Proceedings of the Plasma Space Science Symposium, held at the Catholic University of America, Washington, D.C., June 11–14, 1963.* 1965, IX + 377 pp.
4. Zdeněk Kopal, *An Introduction to the Study of the Moon.* 1966, XII + 464 pp.
5. B. M. McCormac (ed.), *Radiation Trapped in the Earth's Magnetic Field. Proceedings of the Advanced Study Institute, held at the Chr. Michelsen Institute, Bergen, Norway, August 16–September 3, 1965.* 1966, XII + 901 pp.
6. A. B. Underhill, *The Early Type Stars.* 1966, XII + 282 pp.
7. Jean Kovalevsky, *Introduction to Celestial Mechanics.* 1967, VIII + 427 pp.
8. Zdeněk Kopal and Constantine L. Goudas (eds.), *Measure of the Moon. Proceedings of the 2nd International Conference on Selenodesy and Lunar Topography, held in the University of Manchester, England, May 30–June 4, 1966.* 1967, XVIII + 479 pp.
9. J. G. Emming (ed.), *Electromagnetic Radiation in Space. Proceedings of the 3rd ESRO Summer School in Space Physics, held in Alpbach, Austria, from 19 July to 13 August, 1965.* 1968, VIII + 307 pp.
10. R. L. Carovillano, John F. McClay, and Henry R. Radoski (eds.), *Physics of the Magnetosphere, Based upon the Proceedings of the Conference held at Boston College, June 19–28, 1967.* 1968, X + 686 pp.
11. Syun-Ichi Akasofu, *Polar and Magnetospheric Substorms.* 1968, XVIII + 280 pp.
12. Peter M. Millman (ed.), *Meteorite Research. Proceedings of a Symposium on Meteorite Research, held in Vienna, Austria, 7–13 August, 1968.* 1969, XV + 941 pp.
13. Margherita Hack (ed.), *Mass Loss from Stars. Proceedings of the 2nd Trieste Colloquium on Astrophysics, 12–17 September, 1968.* 1969, XII + 345 pp.
14. N. D'Angelo (ed.), *Low-Frequency Waves and Irregularities in the Ionosphere. Proceedings of the 2nd ESRIN-ESLAB Symposium, held in Frascati, Italy, 23–27 September, 1968.* 1969, VII + 218 pp.
15. G. A. Partel (ed.), *Space Engineering. Proceedings of the 2nd International Conference on Space Engineering, held at the Fondazione Giorgio Cini, Isola di San Giorgio, Venice, Italy, May 7–10, 1969.* 1970, XI + 728 pp.
16. S. Fred Singer (ed.), *Manned Laboratories in Space. Second International Orbital Laboratory Symposium.* 1969, XIII + 133 pp.
17. B. M. McCormac (ed.), *Particles and Fields in the Magnetosphere. Symposium Organized by the Summer Advanced Study Institute, held at the University of California, Santa Barbara, Calif., August 4–15, 1969.* 1970, XI + 450 pp.
18. Jean-Claude Pecker, *Experimental Astronomy.* 1970, X + 105 pp.
19. V. Manno and D. E. Page (eds.), *Intercorrelated Satellite Observations related to Solar Events. Proceedings of the 3rd ESLAB/ESRIN Symposium held in Noordwijk, The Netherlands, September 16–19, 1969.* 1970, XVI + 627 pp.
20. L. Mansinha, D. E. Smylie, and A. E. Beck, *Earthquake Displacement Fields and the Rotation of the Earth, A NATO Advanced Study Institute Conference Organized by the Department of Geophysics, University of Western Ontario, London, Canada, June 22–28, 1969.* 1970, XI + 308 pp.
21. Jean-Claude Pecker, *Space Observatories.* 1970, XI + 120 pp.
22. L. N. Mavridis (ed.), *Structure and Evolution of the Galaxy. Proceedings of the NATO Advanced Study Institute, held in Athens, September 8–19, 1969.* 1971, VII + 312 pp.

23. A. Muller (ed.), *The Magellanic Clouds. A European Southern Observatory Presentation: Principal Prospects, Current Observational and Theoretical Approaches, and Prospects for Future Research*, Based on the Symposium on the Magellanic Clouds, held in Santiago de Chile, March 1969, on the Occasion of the Dedication of the European Southern Observatory. 1971, XII + 189 pp.
24. B. M. McCormac (ed.), *The Radiating Atmosphere*. Proceedings of a Symposium Organized by the Summer Advanced Study Institute, held at Queen's University, Kingston, Ontario, August 3–14, 1970. 1971, XI + 455 pp.
25. G. Fiocco (ed.), *Mesospheric Models and Related Experiments*. Proceedings of the 4th ESRIN-ESLAB Symposium, held at Frascati, Italy, July 6–10, 1970. 1971, VIII + 298 pp.
26. I. Atanasijević, *Selected Exercises in Galactic Astronomy*. 1971, XII + 144 pp.
27. C. J. Macris (ed.), *Physics of the Solar Corona*. Proceedings of the NATO Advanced Study Institute on Physics of the Solar Corona, held at Cavouri-Vouliagmeni, Athens, Greece, 6–17 September 1970. 1971, XII + 345 pp.
28. F. Delobeau, *The Environment of the Earth*. 1971, IX + 113 pp.
29. E. R. Dyer (general ed.), *Solar-Terrestrial Physics/1970*. Proceedings of the International Symposium on Solar-Terrestrial Physics, held in Leningrad, U.S.S.R., 12–19 May 1970. 1972, VIII + 938 pp.
30. V. Manno and J. Ring (eds.), *Infrared Detection Techniques for Space Research*. Proceedings of the 5th ESLAB-ESRIN Symposium, held in Noordwijk, The Netherlands, June 8–11, 1971. 1972, XII + 344 pp.
31. M. Lecar (ed.), *Gravitational N-Body Problem*. Proceedings of IAU Colloquium No. 10, held in Cambridge, England, August 12–15, 1970. 1972, XI + 441 pp.
32. B. M. McCormac (ed.), *Earth's Magnetospheric Processes*. Proceedings of a Symposium Organized by the Summer Advanced Study Institute and Ninth ESRO Summer School, held in Cortina, Italy, August 30–September 10, 1971. 1972, VIII + 417 pp.
33. Antonín Rükl, *Maps of Lunar Hemispheres*. 1972, V + 24 pp.
34. V. Kourganoff, *Introduction to the Physics of Stellar Interiors*. 1973, XI + 115 pp.
35. B. M. McCormac (ed.), *Physics and Chemistry of Upper Atmospheres*. Proceedings of a Symposium Organized by the Summer Advanced Study Institute, held at the University of Orléans, France, July 31–August 11, 1972. 1973, VIII + 389 pp.
36. J. D. Fernie (ed.), *Variable Stars in Globular Clusters and in Related Systems*. Proceedings of the IAU Colloquium No. 21, held at the University of Toronto, Toronto, Canada, August 29–31, 1972. 1973, IX + 234 pp.
37. R. J. L. Grard (ed.), *Photon and Particle Interaction with Surfaces in Space*. Proceedings of the 6th ESLAB Symposium, held at Noordwijk, The Netherlands, 26–29 September, 1972. 1973, XV + 577 pp.
38. Werner Israel (ed.), *Relativity, Astrophysics and Cosmology*. Proceedings of the Summer School, held 14–26 August, 1972, at the BANFF Centre, BANFF, Alberta, Canada. 1973, IX + 323 pp.
39. B. D. Tapley and V. Szebehely (eds.), *Recent Advances in Dynamical Astronomy*. Proceedings of the NATO Advanced Study Institute in Dynamical Astronomy, held in Cortina d'Ampezzo, Italy, August 9–12, 1972. 1973, XIII + 468 pp.
40. A. G. W. Cameron (ed.), *Cosmochemistry*. Proceedings of the Symposium on Cosmochemistry, held at the Smithsonian Astrophysical Observatory, Cambridge, Mass., August 14–16, 1972. 1973, X + 173 pp.
41. M. Golay, *Introduction to Astronomical Photometry*. 1974, IX + 364 pp.
42. D. E. Page (ed.), *Correlated Interplanetary and Magnetospheric Observations*. Proceedings of the 7th ESLAB Symposium, held at Saulgau, W. Germany, 22–25 May, 1973. 1974, XIV + 662 pp.
43. Riccardo Giacconi and Herbert Gursky (eds.), *X-Ray Astronomy*. 1974, X + 450 pp.
44. B. M. McCormac (ed.), *Magnetospheric Physics*. Proceedings of the Advanced Summer Institute, held in Sheffield, U.K., August 1973. 1974, VII + 399 pp.
45. C. B. Cosmovici (ed.), *Supernovae and Supernova Remnants*. Proceedings of the International Conference on Supernovae, held in Lecce, Italy, May 7–11, 1973. 1974, XVII + 387 pp.
46. A. P. Mitra, *Ionospheric Effects of Solar Flares*. 1974, XI + 294 pp.
47. S.-I. Akasofu, *Physics of Magnetospheric Substorms*. 1977, XVIII + 599 pp.

48. H. Gursky and R. Ruffini (eds.), *Neutron Stars, Black Holes and Binary X-Ray Sources*. 1975, XII + 441 pp.
49. Z. Švestka and P. Simon (eds.), *Catalog of Solar Particle Events 1955–1969. Prepared under the Auspices of Working Group 2 of the Inter-Union Commission on Solar-Terrestrial Physics.* 1975, IX + 428 pp.
50. Zdeněk Kopal and Robert W. Carder, *Mapping of the Moon*. 1974, VIII + 237 pp.
51. B. M. McCormac (ed.), *Atmospheres of Earth and the Planets. Proceedings of the Summer Advanced Study Institute, held at the University of Liège, Belgium, July 29–August 8, 1974.* 1975, VII + 454 pp.
52. V. Formisano (ed.), *The Magnetospheres of the Earth and Jupiter. Proceedings of the Neil Brice Memorial Symposium, held in Frascati, May 28–June 1, 1974.* 1975, XI + 485 pp.
53. R. Grant Athay, *The Solar Chromosphere and Corona: Quiet Sun*. 1976, XI + 504 pp.
54. C. de Jager and H. Nieuwenhuijzen (eds.), *Image Processing Techniques in Astronomy. Proceedings of a Conference, held in Utrecht on March 25–27, 1975.* XI + 418 pp.
55. N. C. Wickramasinghe and D. J. Morgan (eds.), *Solid State Astrophysics. Proceedings of a Symposium, held at the University College, Cardiff, Wales, 9–12 July 1974.* 1976, XII + 314 pp.
56. John Meaburn, *Detection and Spectrometry of Faint Light*. 1976, IX + 270 pp.
57. K. Knott and B. Battrick (eds.), *The Scientific Satellite Programme during the International Magnetospheric Study. Proceedings of the 10th ESLAB Symposium, held at Vienna, Austria, 10–13 June 1975.* 1976, XV + 464 pp.
58. B. M. McCormac (ed.), *Magnetospheric Particles and Fields. Proceedings of the Summer Advanced Study School, held in Graz, Austria, August 4–15, 1975.* 1976, VII + 331 pp.
59. B. S. P. Shen and M. Merker (eds.), *Spallation Nuclear Reactions and Their Applications*. 1976, VIII + 235 pp.
60. Walter S. Fitch (ed.), *Multiple Periodic Variable Stars. Proceedings of the International Astronomical Union Colloquium No. 29, held at Budapest, Hungary, 1–5 September 1976.* 1976, XIV + 348 pp.
61. J. J. Burger, A. Pedersen, and B. Battrick (eds.), *Atmospheric Physics from Spacelab. Proceedings of the 11th ESLAB Symposium, Organized by the Space Science Department of the European Space Agency, held at Frascati, Italy, 11–14 May 1976.* 1976, XX + 409 pp.
62. J. Derral Mulholland (ed.), *Scientific Applications of Lunar Laser Ranging. Proceedings of a Symposium held in Austin, Tex., U.S.A., 8–10 June, 1976.* 1977, XVII + 302 pp.
63. Giovanni G. Fazio (ed.), *Infrared and Submillimeter Astronomy. Proceedings of a Symposium held in Philadelphia, Penn., U.S.A., 8–10 June, 1976.* 1977, X + 226 pp.
64. C. Jaschek and G. A. Wilkins (eds.), *Compilation, Critical Evaluation and Distribution of Stellar Data. Proceedings of the International Astronomical Union Colloquium No. 35, held at Strasbourg, France, 19–21 August, 1976.* 1977, XIV + 316 pp.
65. M. Friedjung (ed.), *Novae and Related Stars. Proceedings of an International Conference held by the Institut d'Astrophysique, Paris, France, 7–9 September, 1976.* 1977, XIV + 228 pp.
66. David N. Schramm (ed.), *Supernovae. Proceedings of a Special IAU-Session on Supernovae held in Grenoble, France, 1 September, 1976.* 1977, X + 192 pp.
67. Jean Audouze (ed.), *CNO Isotopes in Astrophysics. Proceedings of a Special IAU Session held in Grenoble, France, 30 August, 1976.* 1977, XIII + 195 pp.
68. Z. Kopal, *Dynamics of Close Binary Systems*, XIII + 510 pp.
69. A. Bruzek and C. J. Durrant (eds.), *Illustrated Glossary for Solar and Solar-Terrestrial Physics*. 1977, XVIII + 204 pp.
70. H. van Woerden (ed.), *Topics in Interstellar Matter*. 1977, VIII + 295 pp.
71. M. A. Shea, D. F. Smart, and T. S. Wu (eds.), *Study of Travelling Interplanetary Phenomena*. 1977, XII + 439 pp.
72. V. Szebehely (ed.), *Dynamics of Planets and Satellites and Theories of Their Motion. Proceedings of IAU Colloquium No. 41, held in Cambridge, England, 17–19 August 1976.* 1978, XII + 375 pp.
73. James R. Wertz (ed.), *Spacecraft Attitude Determination and Control*. 1978, XVI + 858 pp.

74. Peter J. Palmadesso and K. Papadopoulos (eds.), *Wave Instabilities in Space Plasmas. Proceedings of a Symposium Organized Within the XIX URSI General Assembly held in Helsinki, Finland, July 31–August 8, 1978.* 1979, VII + 309 pp.
75. Bengt E. Westerlund (ed.), *Stars and Star Systems. Proceedings of the Fourth European Regional Meeting in Astronomy held in Uppsala, Sweden, 7–12 August, 1978.* 1979, XVIII + 264 pp.
76. Cornelis van Schooneveld (ed.), *Image Formation from Coherence Functions in Astronomy. Proceedings of IAU Colloquium No. 49 on the Formation of Images from Spatial Coherence Functions in Astronomy, held at Groningen, The Netherlands, 10–12 August 1978.* 1979, XII + 338 pp.
77. Zdeněk Kopal, *Language of the Stars. A Discourse on the Theory of the Light Changes of Eclipsing Variables.* 1979, VIII + 280 pp.
78. S.-I. Akasofu (ed.), *Dynamics of the Magnetosphere. Proceedings of the A.G.U. Chapman Conference 'Magnetospheric Substorms and Related Plasma Processes' held at Los Alamos Scientific Laboratory, N.M., U.S.A., October 9-13, 1978.* 1980, XII + 658 pp.
79. Paul S. Wesson, *Gravity, Particles, and Astrophysics. A Review of Modern Theories of Gravity and G-variability, and their Relation to Elementary Particle Physics and Astrophysics.* 1980, VIII + 188 pp.
80. Peter A. Shaver (ed.), *Radio Recombination Lines. Proceedings of a Workshop held in Ottawa, Ontario, Canada, August 24-25, 1979.* 1980, X + 284 pp.